Dynamic Memory Management
for Embedded Systems

David Atienza Alonso · Stylianos Mamagkakis
Christophe Poucet · Miguel Peón-Quirós
Alexandros Bartzas · Francky Catthoor
Dimitrios Soudris

Dynamic Memory Management for Embedded Systems

 Springer

David Atienza Alonso
Faculty of Engineering (STI), Embedded
    Systems Laboratory (ESL)
Institute of Electrical Engineering (IEL)
EPFL
Lausanne
Switzerland

Alexandros Bartzas
Dimitrios Soudris
Department of Computer Science, School of
    Electrical and Computer Engineering
National Technical University of Athens
Athens
Greece

Stylianos Mamagkakis
Christophe Poucet
IMEC vzw
Leuven
Belgium

Francky Catthoor
IMEC
Leuven
Belgium

Miguel Peón-Quirós
University Complutense of Madrid
Madrid
Spain

ISBN 978-3-319-10571-0   ISBN 978-3-319-10572-7   (eBook)
DOI 10.1007/978-3-319-10572-7

Library of Congress Control Number: 2014947689

Springer Cham Heidelberg New York Dordrecht London

Printed on acid-free paper

Springer is part of Springer Science+Business Media (www.springer.com)

Preface

Modern embedded systems in mobile and multimedia applications offer a wide range of features. They can also communicate to different devices using different standards which frequently include quite diverse sets of functional and timing requirements. For a good battery life for these devices, they have to be extremely energy efficient. All parts of such a device have to be optimized to reach an acceptable energy efficiency. This book focuses on the dynamic memory management of these devices and how to improve the energy efficiency of the data memory organization in the implementation platforms.

In order to perform a global optimization, we describe a complete, consistent framework and flow for dynamic memory exploration. Such a framework allows the designer to get a complete view enabling to see the global effects of the optimization instead of a narrow view of only local components. This book discusses the elements of the framework and the proposed flow but also the main results and guidelines we have derived from it. We also describe the different steps in the flow with an extensive set of details, such that designers can replicate the main steps and create their own custom flows for a specific target embedded platform. These steps include the effective profiling and analysis of dynamic applications, dynamic data types optimization in multimedia and communication applications, intermediate variable removal to reduce energy and footprint overhead from dynamic applications, dynamic memory management optimization to create memory managers, which effectively deal with internal and external memory fragmentation, and systematic placement of dynamic objects across heterogeneous memory hierarchies.

This book also provides an overview of the state-of-the-art literature in dynamic memory management, intended for data-dominated dynamic applications running on latest generations of portable embedded systems. The material that we present here is based on research at IMEC and its university partners in this area in the period 2001–2013.

Our approach is heavily application-driven which is illustrated by several realistic demonstrators, partly used as red-thread examples in the book. So our approach is illustrated on several representative case studies from the multimedia and communication network application domain. Our target domain consists of embedded systems and especially dynamic systems which deal with medium to large amounts of data. This contains especially multidimensional signal processing (RMSP) algorithms like video and image processing in many different application areas including bio-oriented sensors, graphics, video coding, image recognition, medical image archival, advanced audio and speech encoding, multimedia terminals, artificial vision and graphics processing. But it also includes wired and wireless terminals which handle less regular lists of data. These are present especially in the digital communication network protocol layers.

We therefore expect this book to be of interest in academia, both for the overall description of the exploration methodology and for the detailed descriptions of the main steps. Priority has been placed on issues that in our experience are also crucial to arrive at industrially relevant results. All projects which have driven this research have also been application-driven from the start, and the book is intended to reflect this fact. The real-life applications are described, and the impact of their characteristics on the methodologies and platform components is assessed.

We therefore believe that the book will be of interest as well to senior architecture design engineers and their managers in industry, who wish either to anticipate the evolution of commercially available design concepts over the next few years, or to make use of the concepts in their own research and development.

The material in this book is partly based on work in the context of several European Commission and national research projects. In particular, this work was partially supported by the Spanish Government Research Grant TIC2002/0750 and TIN2008-00508, the EC Marie Curie Fellowship contract HPMT-CT-2000-00031, the European founded programme AMDREL IST-2001-34379 and EC FP7 FET Phidias project (Grant agreement no. 318013), the EC FP7 project 2PARMA-248716, the Swiss NSF Research Grant 20021-109450/1, and the ObeSense (no. 20NA21-143081) RTD project evaluated by the Swiss NSF and funded by Nano-Tera.ch with Swiss Confederation financing.

It has been a pleasure for us to work in this research domain and to cooperate with our project partners and colleagues in the dynamic memory management community. Much of this work has been performed in tight cooperation with many university groups, mainly across Europe. In addition to learning many new things about system synthesis/compilation and related issues, we have also developed close connections with excellent people. Moreover, the pan-European aspect of the projects has allowed us to come in closer contact with research groups with a different background and 'research culture', which has led to very enriching cross-fertilization. This is especially reflected in the many common publications. We want to especially acknowledge the valuable contributions and the excellent cooperation with these groups.

We would like to use this opportunity to thank the many people who have helped us in realizing these results and who have provided contributions in the direct focus of this book, both in IMEC and at other locations. That includes everyone who helped us during the Ph.D. and M.Sc. thesis research.

In particular, we wish to mention: Christos Baloukas, Luca Benini, Geert Deconinck, Vincenzio De Florio, Serge Demeyer, Jose Ignacio Hidalgo, Juan Lanchares, Rudy Lauwereins, Marc Leeman, Jose Mendias, Pol Marchal, Lazaros Papadopoulos, Georgios Pouiklis, Franco Poletti, Jose Risco-Martin, Marijn Temmerman, Antonios Thanalaikis, Liesbet Van der Perre, Karel Van Oudheusden (aka Edgar Daylight), Sven Wuytack, Chantal Ykman-Couvreur, for all their technical feedback during the research that is incorporated in this book. Indirectly, also many others have contributed in some way though and the list would be too long to explicitly mention them all here. But we do acknowledge their support.

We finally hope that the reader will find the book useful and enjoyable, and that the results presented will contribute to the continued progress of the field.

Contents

Chapter 1
Introduction

Embedded systems are computers designed to perform one or more specialized functions and are more cost sensitive than their general purpose counterparts. Today, technology scaling has given them considerable storage and computing power. They use these characteristics to improve the experience and usability or provide new functionality for hundreds of millions of consumer products, such as mobile phones, portable media players, etc. A trend is present towards highly mobile devices, which are able to communicate with each other, access the Internet and deliver rich multimedia content on demand. Embedded systems play a dominant role in this digital vision of ubiquitous computing and are responsible for processing, transferring and storing the heavy data load of all these multimedia and network applications [54]. A resulting major bottleneck is the energy consumption, which affects the operating time, weight and size of the final system (i.e., through the battery). Therefore, it is required to minimize the power consumption while satisfying the memory storage requirements and access requests but also the timing specifications of the various embedded software application tasks that co-exist on the same multi-core hardware platform.

Until the early twentieth century the problem of storing and transferring data to the physical memories in such embedded systems was mostly limited to the management of stack and global data statically allocated at design time [177]. Lately, increased user control (e.g., playing a 3D-game) and interaction with the environment (e.g., switching access points in a wireless network) have increased the unpredictability of the data management needs of each software application. Moreover, in the future this unpredictability will only increase due to the ever increasing functionality and complexity of embedded systems for multimedia and network applications. This is a turning point for the data transfer and storage exploration of these applications, because for an optimal allocation and de-allocation of data it needs to be performed mostly during the run-time phase through the use of heap data (in addition to stack and global data) [11]. *Dynamically-allocated data structures* (termed also Dynamic Data Types or *DDTs* from now on), such as, linked lists, which are present in multimedia user applications (e.g., games, video and music players, etc.) are triggering the allocation, accessing and freeing of their stored elements in quantities and at

© Springer International Publishing Switzerland 2015
D. Atienza Alonso et al., *Dynamic Memory Management for Embedded Systems*,
DOI 10.1007/978-3-319-10572-7_1

time intervals, which cannot be known at design-time[1] and become manifest only at run-time.

As a result, efficient dynamic memory management in embedded systems, especially the ones involving large sets of multimedia data, requires a complete methodology that explores optimizations at two different levels. On the one hand, the selection of suitable DDTs for the running multimedia and network applications executed according to the data memory hierarchy of the final target embedded system. Simple DDTs are dynamic arrays and unidirectional graph-based structures that request memory at run-time (dynamic memory). However, the latest multimedia embedded applications require complex layered DDTs to preserve the relations of stored data, for instance, the objects related to a character in a video game can include a link to the image to display on the screen as well as the amount of life the character still has and its currently available objects to use. Hence, combinations of the two aforementioned basic types are used to construct complex layered DDT structures. The analysis of the design space of these complex data types grows exponentially with the number of layers added to define a DDT. So, a systematic exploration and optimization method for each final embedded memory architecture is required. On the other hand, the dynamic management of the allocation and de-allocation of each memory block associated with each data element stored in the DDTs of an application is performed by a part of the system software, called the *Dynamic Memory Manager (DMM from now on in the book)*. Traditionally, the DMM has been either integrated in the source code of the software application, linked as a separate software library or integrated in the (Real-Time) Operating System. In every case, the DMM is responsible for the allocation and de-allocation of memory blocks. In this context, allocation is the mechanism that searches at run-time inside a pool of free memory blocks (also known as the *heap*) for a block big enough to satisfy the request of a given application. De-allocation is the mechanism that returns this block back to the pool of memory blocks, when it is no longer needed. The blocks are requested and returned in any order, thus creating "holes" among used blocks [171]. These holes are known as *memory fragmentation*, which forms an important criterion that determines the global quality of the DMM, because a possibility exists that the system runs out of memory, and thus crashes, if fragmentation levels rise too high. Besides memory fragmentation other criteria are present regarding the DMM cost efficiency, which are related to energy consumption, memory footprint, memory accesses and performance.

In this book, we thoroughly investigate the design of the dynamic memory management subsystem for cost-sensitive embedded systems with hard and soft real-time requirements. We discuss how to design suitable multi-layer DDTs for the final data memory hierarchy of the target embedded system, and how to create a low fragmentation, cost efficient DMM out of parameterizable components for the particular memory allocation and de-allocation patterns of the different types of embedded multimedia and network applications. The design methodology is based on propagating constraints among design decisions and parameterizing according to application spe-

---

[1] Throughout the text "design-time" and "compile-time" are used interchangeably.

cific data access and storage behavior. We show a systematic approach of performing DDT and DMM design exploration using the academic version of the MATISSE prototype tool framework which we have developed. Finally, we assess the effectiveness of the proposed methodology to optimize several real-life cases of multimedia and network applications in the latest embedded systems.

1.1 Embedded Systems

Modern embedded systems are a combination of software and hardware, which performs a single or a few dedicated functions. In the context of this book, we focus on embedded systems that perform more than one function and more specifically, on the category of *nomadic embedded systems*. These are mobile devices that have multimedia or network functionality (or both) and provide to the end user the ease of rich multimedia content anytime, anywhere. To provide this functionality, they rely on hardware platforms, which integrate a number of processing elements, communication infrastructure and data memory hierarchy on a single chip (or multiple chips on a board). The processing elements in embedded platforms became massively programmable since early 2000, which means that they can perform more than one function according to the embedded software that accompanies them [36]. We classify embedded software as *application software* and *system software*, as can

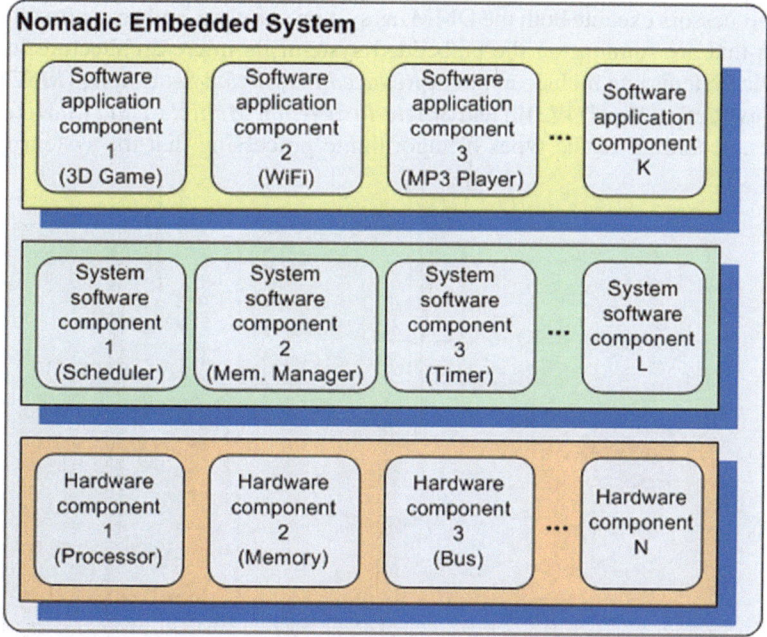

Fig. 1.1 Hardware and software components in an embedded system

be seen in Fig. 1.1. In particular, application software provides relevant services to the end user and the system software interfaces with hardware to provide the necessary services for application software. For example, on the one hand we consider a 3D-graphics rendering software as a software application and on the other hand we consider DMM software as system software. The embedded system designers are responsible for the design, integration and optimization of the hardware, application software and system software components. In the following subsections, we will give an overview of the component subset that we are considering in the context of this book, which are applicable to a large set of processor-based portable embedded systems.

1.1.1 Hardware Platform

Many different platform designs exist according to the number, type and interconnection of the hardware IP components used. In this book, we target platforms that can be represented by the generic design template seen in Fig. 1.2. This template consists of programmable microprocessor, communication network, Direct Memory Access (DMA), Memory Management Unit (MMU) and multilevel data memory hierarchy [36, 63].

The programmable microprocessor cores form the key element that execute the instructions and, thus, they orchestrate the overall behavior of the final embedded system perceived by the user [74]. It is important to note that the programmable microprocessors execute both the DMM, as system software, and any software applications that are running on the embedded system. Its micro-architecture in latest nomadic systems can include a mix of *Reduced Instruction Set Coding (RISC)*, *Very Long Instruction Word (VLIW)* and *Single-Instruction Multiple-Data (SIMD)* architectures, according to the types of algorithmic processing that the system mainly

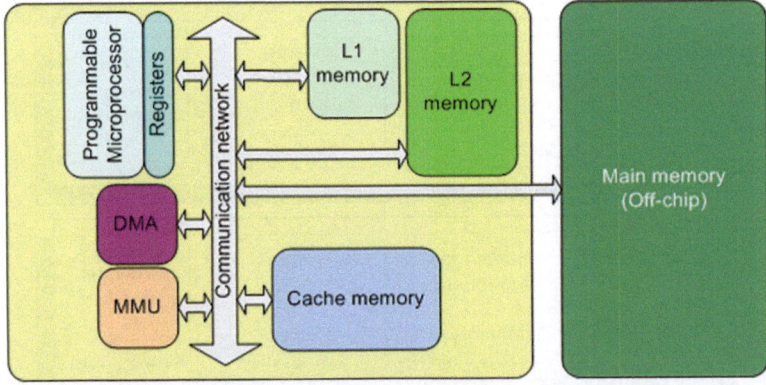

Fig. 1.2 Generic platform design template with multilevel data memory hierarchy

executes. In the rest of the book we consider a multi-core platform with the presence of at least a RISC-type processor for orchestrating the background data management including the DMM aspects.

The multilevel data memory hierarchy represents the physical memories that are used in the embedded system to store data. Also memory is present for storing the instructions, but the optimization of the management of such memory is out of the context of this book. As can be seen in Fig. 1.2, all the data is stored off-chip in the main memory (with storage sizes between a few and several GBytes), while a smaller portion of the data (usually less than a few MBytes) is stored in the on-chip memory. Today's platforms usually have two or three levels of on-chip memory (L1 and L2 and possibly L3 with up to 32–128 KBytes and several MBytes, respectively) [74]. The type of memories that are in the center of the methodologies addressed in this book are on-chip scratchpad memories or software-controlled caches, which means that software decides which data will be written on or read from them [168]. This type contrasts with hardware-controlled on-chip memories or caches, which include hardware logic that decides autonomously which data should be stored on them at each moment in time during the execution and, thus, replaces the blocks stored on-chip accordingly [74]. Note that the described methodologies are not blocked by the existence of hardware-controlled memories, nevertheless they can take full advantage of scratchpad or software-controlled memories and provide further optimizations. Finally, besides programmability, the proposed methodologies are initially independent of the memory type used (e.g., SRAM, DRAM, SDRAM, Flash), as the inclusion of one or another translates only in different cost functions for each dynamic memory management operation. Hence, the final dynamic memory management solutions can vary, but the exploration and optimizations discussed in this book remain valid. Still, for quantitative case studies and comparisons, all energy measurements for on-chip memories will be performed with SRAM models [77, 80].

The communication network ensures the data transfer from the data memory hierarchy to the microprocessor registers and vice versa. Additionally, the communication network connects the individual hardware components of the embedded system and makes possible the transfer of data between the levels of the data memory hierarchy without using the host microprocessor (via the DMA) [36]. According to the complexity of the platform template, the communication network can be implemented with a variety of designs like a simpler bus or a more complex *Network-on-Chip (NoC)* [55]. The methodologies that are described here aim to reduce the memory accesses (see Sect. 1.3.3) rather than the cycles needed for performing such a memory access [77], thus they can be considered independent of the communication network. In the context of this book we use a simple shared bus between all the system components for illustration purposes.

Finally, the DMA gives the opportunity to transfer data between the data memory hierarchy levels without interrupting the microprocessor, which is responsible for data transfers when a DMA is absent or when it is not programmed to move a particular set of data [36]. The DMA programming can be performed by the embedded software designer with specialized instructions inserted directly in the source code of a software application. In addition, in some processing architectures, a *Mem-*

ory Management Unit (MMU) exists to translate the virtual memory addresses to physical memory ones, which can be mapped directly onto the memories of the embedded system [74]. However, in this book all references to memory addresses are assumed to correspond to physical memory addresses. Without loss of generality, we could also have included a virtual translation layer. In that case an MMU would be necessary and this could be used in a seamless integrated way with respect to the proposed optimization methodologies for dynamic memory. In contrast, the DMA is assumed to be present in the target architecture when software-controlled memories or scratchpad memories are used, which is typical in multimedia embedded system-on-chip designs (e.g., Samsung Exynos 4 Quad [141], Freescale i.MX287 [62], TI Omap [161], etc.). Nevertheless, the usage of a DMA is not really needed to enable the introduced proposed optimizations, rather its presence increases system performance complimentary to the DMM and only affects the cost function of memory transfers for scratchpad memories [135].

Finally, the current wave of the *Multiprocessor Systems-on-Chip (MPSoCs)* involves its own design of data memory hierarchies (a template example can be seen in Fig. 1.3) [20]. The main differences (from the hardware viewpoint) with respect to the generic template architecture previously presented relates to the overall number of microprocessors, memories and layers in the data memory hierarchy to optimally transfer the shared data between the processors, as well as the communication network that becomes more complex due to the large number of system components. In this case, shared data and distributed memory address spaces comprise an even more complex data memory hierarchy, which has a significant effect on the cost factors regarding memory management. The methodologies introduced

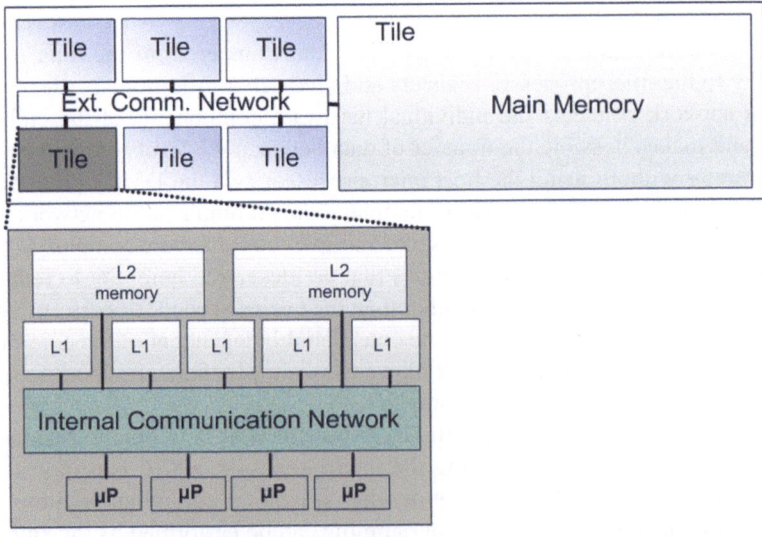

Fig. 1.3 Multi-processor platform design template

in this book are compatible with the data memory hierarchies in MPSoC platforms, but do not consider the additional requirements of dynamic memory shared between the multiple processors present in multi-core architectures, with distributed memory hierarchies and the derived new cost factors. Also, coherency and atomic data access mechanisms for shared dynamic memory in presence of multiple concurrently executing threads are not discussed as such here. We believe though that the material present in the methodologies and techniques proposed by us can become an integral part of the extended methodologies required for these multi-core platforms. They only needed to be augmented with appropriate cache coherency mechanisms that exist in the state-of-the-art literature.

1.1.2 Embedded Software Applications

In embedded systems, embedded software diversity is growing very fast. Since hardware becomes more programmable, multiple software applications can run on the same embedded system. In this context, we define as embedded software (from now on, we will simply refer to them as *applications* in the text) the different programs, which are written to provide the different user services in the final device, such as, games, office applications or music and video players. It is important to note that software applications receive input from the end user of the embedded system. As a consequence, the actions of the user have great impact on the control flow and memory utilization of the algorithms implemented in the running applications; thus, the execution of the applications becomes dynamic regarding the amount of resources used and event-driven.

In the context of this book, we focus mainly on the multimedia and network application domains for nomadic embedded systems, which on top of increased dynamic behavior also show increased data transfer and storage activity. These two particular characteristics make the targeted domains particularly suitable for the dynamic memory management optimizations introduced here. In the two following subsections, we give an overview of the two main target domains.

1.1.2.1 Multimedia Applications

Multimedia embedded applications have experienced recently a very fast growth in their variety, complexity and functionality. Typical and successful examples of such type of applications are video and music players [59, 117, 124], three-dimensional games [45, 104, 122] or *Voice-Over-IP* (*VOIP*) [121] applications. Nowadays, these new multimedia applications (e.g., MPEG4) include complex algorithms, coded in object-oriented languages (e.g., C++ or JAVA), and have largely increased the demands of memory and performance of traditional embedded software. Moreover, their graphic and data processing requirements are unknown at compile time. This is due to the fact that they depend on the specific input (e.g., resolution of streaming video) and user behavior (e.g., full screen or small window size), which can result in unaffordable high cost and power consumption embedded systems without extensive optimizations for the target embedded platform. For instance, in *three-dimensional* (*3D*) object rendering applications, such as, MPEG-21 or in applications with adaptable *Quality of Service* (*QoS*) features [136], variations of two orders of magnitude can be observed according to the extreme cases in the number of objects (i.e., one object at low resolution or 15 objects at high resolution per frame, respectively) that need to be stored in memory and rendered per frame. Thus, their porting to embedded systems is a very time-consuming and prone process.

In this book, we present in detail the results of the application of the proposed methodology to optimize the dynamic memory management subsystem to three different real multimedia applications. Our first case study is a new 3D-video rendering application based on scalable meshes [104] that adapt the quality of each object displayed on the screen according to the position of the user. The objects are internally represented by vertices and faces (or triangles) that need to be dynamically managed in the meshing algorithm and corresponding complex data structure due to the uncertainty at compile time of the features of the objects to render. The original DDT consists of a dynamically created tree where vertices and faces are stored separately. This data structure needs to be traversed according to different access patterns (i.e., the different rendering phases [104]) to render them onto the screen. First, the vertices are traversed during the first three phases of the whole visualization process. Then, the faces are processed in the final three phases [104] of the rendering process to show the objects with the appropriate resolution on the screen. The dynamism in this application appears in all the different phases of the rendering process, since they are greatly affected by the features of the stream of input images.

The second case study used in this book is a crucial module from a 3D-image reconstruction application [138] that matches corners from two subsequent frames and thus creates the mathematical abstraction to be used in the whole application (see [138] for references to the full code of the algorithm with 1.75 million lines of high level C++). The operations done on the images are particularly memory intensive, e.g. each image with a resolution of 640 × 480 uses over 1Mb, and the accesses to the images are randomized. Thus, classic image access optimizations as row-dominated accesses versus column-wise accesses are not relevant and the overall amount of energy dissipated in the system is unacceptable in case of worst-case assumptions to apply static (i.e., compile-time) optimization methodologies [35, 36].

Finally, the third case study is an interactive 3D-game that integrates virtual-reality objects in scenarios offered by the real world thanks to a frame grabbing device [52, 60]. In real-time, images are taken from our everyday world and obstacles (i.e., walls) are detected on them. Then, in the scenario of detected walls, additional virtual objects (i.e., balls) are generated and interact with them. The software module studied is the game controller, which handles the actions of the balls (e.g., movements or impacts) and the interaction between the scenario and the user. Three relevant DDTs are present in this module. First, VWalls is the list of vertical walls detected in the input images. Second, HWalls is the list of horizontal walls. Third, Balls is the list of current balls in the system. These DDTs were originally implemented using variations of single-linked lists and they require to be optimized to fit in current embedded handheld devices, while respecting the real-time requirements of the frame grabbing device.

The three previously explained applications will be the basis for our discussions in Chaps. 2 and 7 about the features of the latest and forthcoming embedded multimedia and network applications, and they will illustrate the optimization results regarding the dynamic memory management subsystem (both for DDTs and DMM design) in real-life applications, using the overall design and optimization methodology presented by us.

1.1.2.2 Wireless Network Applications

In the last years networks have become ubiquitous. Modern portable devices are expected to access the Internet (e.g., 3G mobile phones) and communicate with each other in a wireless way (e.g., tablets with 802.11b/g). In order to provide the desired Quality of Experience to the user, these systems have to respond to the dynamic changes of the environment (i.e., network traffic) and the actions of the user (i.e., visiting web pages) as fast as possible. A large variety of software applications (written in many programming languages) implements these mobile communications on an embedded system.

In the context of this book, we use the term wireless network applications to refer to the C or C++ source code that is running on an embedded system and is responsible for the connection and communication of this embedded system with another embedded system (Access/Mobile Point) or with a server. The basis of the networking theory for wireless embedded systems is derived from the networking theory for general purpose computing systems [159], which can be split into 7 layers according to the Open Systems Interconnection (OSI) Basic Reference Model, as can be seen in Fig. 1.4. The introduced methodologies are relevant for the Data link, Network, Transport, Session, Presentation and Application layers (i.e., every layer except the Physical layer).

The memory access and storage behavior of these wireless network applications is very dynamic, which means that their resources vary at run-time in a way not predictable at design-time. They have heavy control flow and the loops are focused on tasks repeated for every network packet received and transmitted. The data storage is dominated by the network packets that are stored in memory to be processed in each layer of the OSI model. The processing is then responsible for most of the data accesses that take place. They are focused on the headers of the network packets as they read and write on its fields. The main case study we have used is a Deficit Round Robin (DRR) scheduler. Today's frame based networks are using data packets of fixed or variable length and memory is allocated once a data packet enters a layer, then its header fields are accessed and finally, the memory is freed once it exits a layer and is forwarded. In general, the size of a packet and the timing of its transmission/reception cannot be predicted, because it is affected by the user actions and the environment.

OSI Model			
	Data unit	**Layer**	**Function**
Host layers	Data	7. Application	Network process to application
		6. Presentation	Data representation and encryption
		5. Session	Interhost communication
	Segments	4. Transport	End-to-end connections and reliability (TCP)
Media layers	Packets	3. Network	Path determination and logical addressing (IP)
	Frames	2. Data link	Physical addressing (MAC & LLC)
	Bits	1. Physical	Media, signal and binary transmission

Fig. 1.4   OSI model with seven abstraction layers for network protocols

Nevertheless, some safe conclusions can be drawn about the dynamic access behavior and memory storage resource needs of wireless network applications, which will be discussed in Chaps. 2 and 7.

1.1.3 System Software

The system software refers to the software which manages and controls the hardware so that application software can perform a certain task [74]. Contrary to application software, the functionality of the system software is transparent to the user. Popular types of system software include middleware components (e.g., CORBA), operating system components (e.g., Linux), device drivers and firmware components. In modern embedded systems, with multiple software applications sharing a variety of hardware resources, the use of system software is required to ease and increase the efficiency of the resource management. In the context of this book, we focus on the memory management system software, which manages the data transfer and storage of software applications on the data memory hierarchy.

As can be seen in Fig. 1.5, the memory manager mediates between the software application and the actual storage on the physical memories. The memory manager itself is actually software which interfaces via function calls (or methods) to the source code of the software applications and is usually given in the form of a library file, which is compiled and linked with the software application. Usually it is very closely associated with a particular operating system and changes according to the type and distribution of the operating system. The memory manager is responsible for every allocation and de-allocation of memory in the embedded system. Further

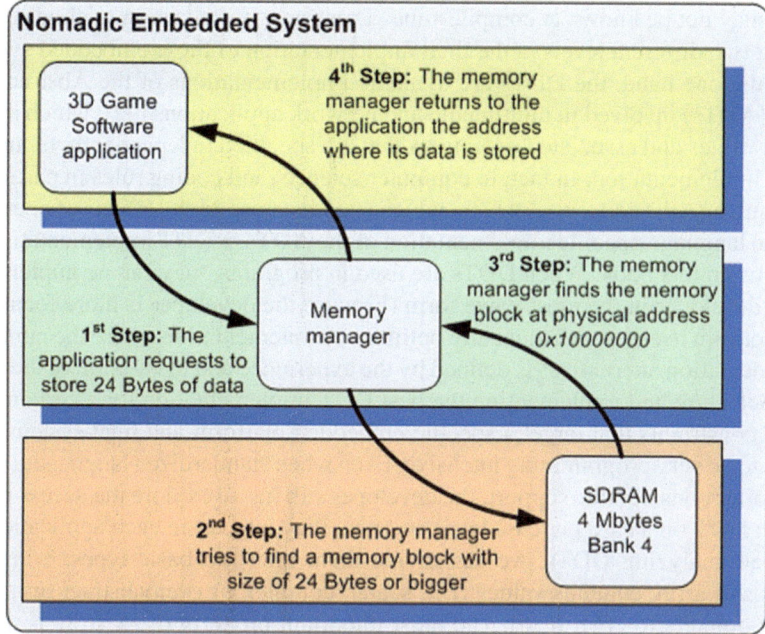

Fig. 1.5 The memory manager mediates between software application and hardware memory for memory allocations and de-allocations

in the book we will distinguish between static and dynamic memory management referring to the management of the stack and global data on the one hand and to the management of the heap data on the other hand. As already indicated, the task of the dynamic version of the memory manager will be termed DMM.

1.2 Problem Definition

The problem that this book addresses is related to the dynamic nature of new embedded multimedia and network systems. In this type of systems, software applications allocate, access and de-allocate data in the memory of the embedded system at times and quantities that are not known during its design phase. Thus, more traditional memory optimization methods for embedded systems [36], which statically or pseudo-statically (i.e., limited number of execution options or scenarios) analyze access patterns and sizes of data transfers between the data memory hierarchy layers in consumer applications in order to select the best data placement, cannot be applied. Moreover, efficient DMM of dynamic embedded applications for the latest multimedia and network applications implies new optimization methods that can analyze and optimally handle a large variation of possible input set characteristics at run-time, which may not be known at compile-time. These new optimization methods need to work at two different levels in the final implementation of those embedded systems.

On the one hand, the DDTs are dynamic implementations of the Abstract Data Types (ADTs) involved in multimedia and network applications [42], which are sets of data values and associated operations that are specified independently of any particular implementation. In fact, in computer science (and coding rules in particular), we mainly use the concept of ADTs, which stresses more on the effects of operations than the language-specific implementation of the ADTs or DDT implementation that we focus on this book. When DDTs are used in programs, they can be implemented by the developer in the most naive form (because the developer is more focused on the algorithm itself) or in a manually optimized implementation where the number of implementation alternatives is defined by the experience and inspiration of the developer. Selecting and implementing the best DDT implementation for a certain set of design constraints that target a specific embedded platform and final system, often proves to be very programming intensive. Even when standardized languages (if used at all) offer considerable support, the developer still has to explore the access pattern of each DDT on a case per case basis and how they interact in each application.

When analyzing DDTs, we can divide them into two basic types: arrays and graphs. An array contains values (i.e., scalar, complex or even abstract ones when storing pointers to other data). The most important property of an array is that the entire structure is allocated on one shoot, but the memory can be allocated at compile time (static memory) or at run time (dynamic memory). On the other hand, graphs consist of multiple nodes that can store a scalar or complex value each. Furthermore, it contains at least one pointer to another similar memory location (that can be NULL). The combination of all connected nodes forms the complete graph. Contrary to arrays, nodes in a graph are allocated and freed incrementally: when more data needs to be stored, additional nodes are allocated. In most cases, two special cases of graphs are used: trees and lists. In a tree, each node has one *parent* node and at least one *child* node. Exceptions are the *root* node that has no parent and the *leaf* nodes that have no children.

Complex layered data types are combinations of the two aforementioned basic types. In a typical example, the overhead memory in linked lists (i.e., the pointers to the next and previous nodes) are amortized by allocating memory for several elements at once. In this case, the first layer is a linked list and the second one is an array. As a result of the possible combinations of basic types in complex layered structures, the search space of these complex data types grows exponentially and a systematic exploration and construction method as we present in this book becomes a must.

On the other hand, the DMM memory manager is responsible for the management of the system's memory and to provide a valid memory address in order to satisfy the requests of the software application. Therefore, the memory manager constitutes an intermediate design abstraction layer between the software application and the hardware memory of the embedded system. The way that the data is allocated and assigned to the individual memories of the data memory hierarchy will have a significant impact on the performance, energy consumption, use of the interconnect bandwidth and the size of the memories needed. More specifically, the moment that the software application will ask memory from the system (in order to store its data) is not known beforehand by the embedded system designer (who is also the designer of the memory manager). This happens because modern multimedia and network applications exhibit a high degree of interaction with the user of the application, whose decisions cannot be fully anticipated at design-time and thus vary the needs of the application for memory space. For example, the user selects at an arbitrary moment to sent an email or to browse a webpage. It is obvious that these two different user actions, which are considered as input for the embedded system, can be made in any order and in any time instance. Nevertheless, the user expects from the software application, and thus from the dynamic memory manager, to react and satisfy immediately in an optimal way the requests, namely, without wasting the limited available memory and processing resources.

Additionally, other sources of uncertainty (besides the user actions) are the great abilities of modern software applications to adapt to their environment. We define as environment anything that is not decided by the user or the software application itself. For example, a network may become congested, thus forcing the wireless network application to delay the transmission of data and storing it for a longer time period in the memory. In this situation, the dynamic memory manager does not know beforehand the moment that the application will be able to send the data and thus will not require to store them any longer, which in turn will enable the dynamic memory manager to free the memory where the data was stored.

Finally, the characteristics of user decisions or the situation of the environment affect the size of memory that the software application requests each time from the dynamic memory manager. A typical example is the difference between the memory requested during loading a webpage with rich multimedia effects and during loading a webpage with just text. It is obvious that the software application that loads the webpage will request much more memory in the first case because it will have to deal with bigger quantities of data. Nevertheless, the DMM does not know in advance which page will be selected by the user and thus the needs of the software application, which request the memory.

To sum up, requests for memory are performed at unknown time instances and for unknown memory sizes. Additionally, it is unknown for how much time the memory will be reserved until the data is de-allocated. Therefore, the designer of the memory management subsystem does not have the fixed specifications at design-time, which would enable the straightforward solution of the memory management problem. The solution of the memory management problem should be dynamic rather than static in order to be able to adjust the amount of memory used by the data structures of the application and the final management mechanism of the dynamically allocated memory, according to needs for memory during the execution of the software application, as can be seen in Fig. 1.6. By dynamically adapting the data structures, we can improve both the overall performance and the energy requirement for the executed application tasks.

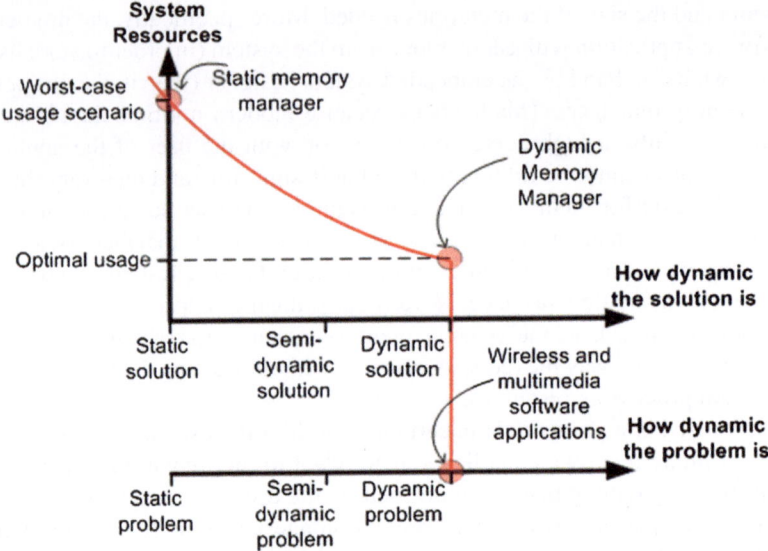

Fig. 1.6 Static memory management solutions fit for static data storage needs and dynamic memory management solutions fit for dynamic data storage needs

1.2.1 Shortcomings of Static Solutions

Static memory management solutions are worst-case solutions for embedded applications that exhibit dynamic access and storage behavior. More specifically, we define as static data structures the solutions that allocate their memory when one user application starts its execution, and de-allocate it only when the application is terminated. To this end, static data structures must allocate their maximum possible size, which is performed by the compiler using the stack and main memory. In contrast, DDTs allocate their data in a reserved memory region for dynamically allocated data, called *heap*, and can vary their allocated sizes at run-time to better adjust their memory utilization to their actual requirements at each moment in time. Thus, the inefficiency of static memory solutions for dynamic applications contrasts sharply with the efficient use of dynamic memory of DDT and DMM solutions.

In the case of the static memory solutions, the worst use case scenario must be evaluated (i.e., the scenario that the software application needs to store the maximum amount of data). After this worst-case scenario is calculated, the static memory manager of the system will allocate the maximum amount of memory at the beginning of the execution without de-allocating it later. Therefore, the static memory management solutions confront the following four challenges:

1. Every time that data needs to be stored, the worst-case size is assumed for every data storage request. This problem takes big proportions in software applications

(e.g., like network applications) that make thousands of variable-sized data storage requests during a limited time period.

2. The memory blocks, which store data that is no longer used, cannot be de-allocated at the execution time of the software application. Therefore, these memory blocks cannot be used to satisfy data storage requests of the same or another application running concurrently on the same embedded system. Especially, for multimedia and wireless network applications, which do not store data for an extended time period rather they do it in phases, the aforementioned restriction is very unfavorable for the final embedded system design.

3. It is very difficult, or even impossible sometimes, to estimate the worst-case resource usage scenario. In the case that the actual worst-case scenario for a certain input of the application is better than the calculated worst-case scenario, then memory resources are wasted. Conversely, if a software application requests memory to store data above the estimated worst-case threshold at design time, the system would crash. Hence, we cannot any longer provide safe design-time bounds for mission-critical applications that require guarantees. The only safe solution is to allow the alternative of a dynamic allocation (discussed above) which incorporates the limited size of the actually available physical memory while also supporting the data needed to correctly execute the application tasks.

4. The source code of the software application must be rewritten in order to transform all the dynamic data storage requests of dynamic data structures (e.g., like linked lists) to worst-case data storage requests of static arrays, so that the use of a dynamic memory manager is avoided. The source-to-source transformations needed for modern multimedia and network applications are very time consuming since they consist of thousands of lines of complex source code, which is full of algorithmic interdependencies between the various subroutines and respective data structures.

To sum up, static memory management solutions are typically not used anymore in modern multimedia and network applications, mostly due to the first two problems that have a significant negative impact on the use of system resources. Therefore, at least partially, dynamic memory management solutions must be evaluated and implemented for these new dynamic software applications when executed in latest and forthcoming nomadic embedded systems.

1.3 Metrics and Cost Factors

Optimization is especially important in embedded systems in order to decrease cost, to increase the satisfaction of user experience and, most importantly, to meet very tight design constraints in energy consumption, memory footprint and interconnect bandwidth usage. The quality of a DMM solution is characterized by a set of metrics and cost factors. The metrics are related to the efficiency of the used DDTs and the DMM functionality, namely, which data allocation requests exist in the application and how they are served by the systems. The cost factors relate the positive or negative contributions of the DMM to the cost of each device that hosts it. In the next subsections it will become obvious to the reader that an optimized design can significantly increase the efficiency of the DMM and decrease any negative impact of the DMM to the cost of the embedded system.

1.3.1 Memory Fragmentation

The most important metric representing the efficiency of a DMM solution is the memory fragmentation. This fragmentation is present in the system memory, which the DMM has under its control. As a result of a too high fragmentation, it cannot use that space any longer to satisfy all data storage requests. The total memory fragmentation is split into internal and external memory fragmentation (see Fig. 1.7).

When the application asks for memory to provide new storage space for elements in the DDTs of the application, the DMM returns a specific memory address where the application can write the data it needs to store. More specifically, in the address that is returned resides a free memory block, which can be used for as long as the

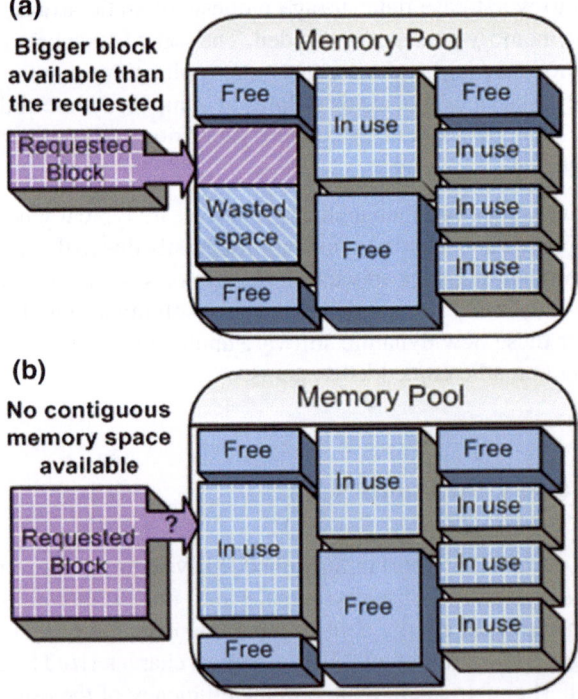

Fig. 1.7 a Internal and **b** external memory fragmentation

application needs to store its data there. During the period that the data is useful to the application (i.e., it is accessed), the memory blocked remains allocated and the DMM cannot use it to satisfy another memory request of the application. When the data is not useful anymore, then the memory block is de-allocated and the DMM can reuse it in order to satisfy the next data storage request.

When memory is allocated, it is said that the DMM returns a block of memory to the application. When memory is de-allocated, it is said that the DMM frees the memory block that was allocated. As mentioned before, the requests of the software application for allocation and de-allocation of memory can happen for any size, at any time and in any order. This means that after some time of DMM operations, the memory consists of free and used memory blocks of various sizes.

Internal fragmentation happens when the DMM allocates a block of memory, which is bigger than the size of the data storage request of the software application. While the memory block remains allocated, its extra memory space which is not used to a store data, remains also allocated and cannot be used to store other data. External fragmentation happens when one big data storage request cannot be satisfied while enough smaller free memory blocks exist, which could be coalesced to satisfy the request. This failure is attributed either to the fact that these smaller blocks do not lie in consecutive memory addresses or that no efficient coalescing algorithms are present inside the DMM. Therefore, the DMM cannot use the memory space of some blocks (which lie between allocated blocks or are simply too small) besides the fact that they themselves are free. The total memory fragmentation is the sum of external and internal fragmentation and is defined as the memory that is managed by the DMM to the data size that is requested by the application minus one. The designer of the embedded system has to design a DMM in such a way that the total memory fragmentation is minimized. Only minimal fragmentation ensures the correct and efficient operation of the system.

1.3.2 Memory Footprint

We define as memory footprint the total memory size that needs to be managed by a DMM in order to fulfill the data storage needs of the DDTs for a given application with a specific input. The memory footprint is directly correlated to the memory fragmentation and can be defined as the sum of: (i) the internal fragmentation, (ii) the external fragmentation, (iii) the memory needed to accommodate the internal data structures of the DMM and (iv) the data size that the DDTs of the application need to store.

While the data size storage request cannot be controlled by the DMM designer, a successful design can minimize the internal fragmentation, the external fragmentation and the memory needed by the data structures of the application and the internal ones of the DMM itself. Therefore, the minimization of the memory footprint that is managed by a DMM can help to significantly reduce the physical memory that needs to be integrated in the platform of the final embedded system.

The minimization of the physical memory of the embedded system is very important because it influences directly the cost of the device (i.e., smaller memories are cheaper). Especially, in the case of embedded systems that realize multimedia and network applications, which make mass consumer electronics a reality, one small reduction in memory cost is multiplied by the millions of units that are manufactured. Additionally, decreasing the physical memory size can have positive effects on the chip size, since memories take up most of the area on a chip [74]. Minimizing the area allows either the shrinking of the final device (i.e., making it more attractive for the consumer) or the integration of more functionality in a device with the same size.

1.3.3 Memory Access Count

We define as memory access count the sum of: (i) the number of accesses that is needed by the DMM to allocate and de-allocate memory and (ii) the number of memory accesses needed by the dynamic data types to read and write on the stored data. As analyzed in Chaps. 3 and 4, on the one hand some applications store data only a couple of times and then access them heavily at run-time and other applications store data millions of times and rarely access it afterwards. Depending on the ratio of storing versus accessing the data, design optimizations on the DMM, on the DDTs or on both are needed in order to decrease the memory accesses in the embedded system.

An improved design can decrease dramatically the number of memory accesses, which enables the minimization of the bandwidth needed and thus the on-chip interconnect can be used more efficiently. Bandwidth requirement is an important factor of the embedded system performance, since it commonly becomes a bottleneck, especially in the case where multiple software applications are executed concurrently. Additionally, the on-chip interconnect takes up a considerable amount of chip area, which can be subsequently minimized indirectly with the memory access minimization.

1.3.4 Application Task-Level Performance

We define as performance the time needed by one multimedia or network application to handle a particular task set or workload and to store and access its data dynamically via the selected DDTs and underlying DMM to perform the final (de)allocation on the physical memory space for dynamic memory or heap space. Because multimedia and network applications have an especially intensive data access and storage profile, the speed of the DMM and DDTs are very important and affects significantly the performance of the whole application. In fact, inappropriate memory management can comprise a larger portion of the execution time of the application than data computation itself [36]. The main reason is that the memory accesses (performed by the DDTs) and the final memory assignment of dynamic memory (performed by the DMM) play a very significant part in the performance of the software application because it determines how close the data is to the processor and thus the latency of a memory access. As a result, according to the set of applications to be executed in the final embedded system, the design of a DMM and DDTs can be significantly tuned to reach significant speed-ups with respect to general purpose solutions, while reaching the same or lower fragmentation levels, as we show in this book.

In addition, improving the performance of the DMM and DDTs can decrease the cost of an embedded system by decreasing the needs for higher speed microprocessors and other hardware components, while still meeting the deadlines and the design specifications. A system without these optimizations would require a higher performance, more expensive microprocessor in order to be able to handle the same workload in the same time frame. In case it is decided not to use another system architecture, despite the DMM and DDT optimizations, the same hardware resources can be used to provide higher QoS. Alternatively, the resources freed by the optimizations can be reused to provide additional functionality in the existing applications or enable the embedded system designer to execute more applications in parallel.

1.3.5 Overall Energy Consumption

We define as energy consumption for a given task set the energy that is consumed by the physical memories and is attributed to the activity of the DMM or the DDTs. The energy consumption is affected by three factors: (i) the memory footprint of each physical memory controlled by the DMM (for a given technology node), (ii) the total amount of accesses to each physical memory required for the task set and (iii) the assignment of the data in the memory hierarchy. As already mentioned all these factors are responsibilities of the DMM. Note that energy and (instantaneous) power consumption are quite different aspects, and for embedded systems typically the former is the most important.

More specifically, each access in the memory organisation consumes energy. As can be seen in Fig. 1.8, accesses to the memory blocks that are assigned to on-chip memories consume far less memory than the accesses to memory blocks that are

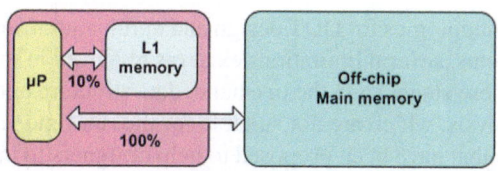

Fig. 1.8 Memories which are on-chip consume less energy per access compared to off-chip memories

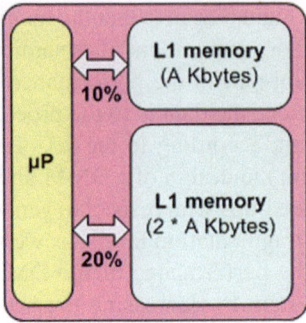

Fig. 1.9 Memories which are smaller in size consume less energy per access compared to memories with bigger size

assigned to off-chip memories. Additionally, bigger physical memory size means memories with bigger silicon footprint, which consume more energy per access compared to smaller memories (as can be seen in Fig. 1.9).

As can be seen in the function below, the power per access that is consumed by the memory is affected directly via the effective capacitance C_{eff} (i.e., bigger memories have bigger C_{eff}). The memory accesses affect the power consumption via the real access frequency f_{access}. Finally, the drain voltage V_{dd} affects as well the power consumption but the V_{dd} itself cannot be controlled via methodologies presented in this book.

$$P = \frac{1}{2} \times C_{eff} \times (V_{dd})^2 \times f_{access} \tag{1.1}$$

The minimization of the energy consumption via the optimized DMM design enables either an increase in operating time and autonomy of the embedded system or reduction in weight, size and cost (i.e., by using a smaller battery). Finally, energy consumption affects the heat dissipation of the memory (via the power consumption), thus an optimized DMM needs a less costly heat sink.

1.4 Overview of the Proposed Approach

In the existing methodologies for DDT design and optimization, and DMM solutions for embedded systems, several limitations exist (as reviewed in the respective chapters dealing with these stages). On the one hand, data structure optimizations mainly focus on static analysis, which are not suitable for the run-time variations of DDTs. Also, the solutions that have been proposed to help designers to implement complex multi-layer DDTs mainly focus on providing general interfaces according to abstract data types specifications, and implementations without taking into consideration the specific underlying data memory hierarchy, which does not enable easy and suitable customization of the DDTs for multi-objective optimization metrics (e.g., low-power, performance and memory footprint), as embedded systems for new multimedia and network applications require.

On the other hand, the existing approaches related to DMM design mainly propose solutions that try to satisfy a wide range of run-time data storage needs of various types of applications at the same time. Thus, they propose one-size-fits-all solutions without focusing on the specific features of multimedia embedded applications (e.g., games, routing protocols, video processing, etc.). Moreover, these off-the-shelf DMMs do not use any platform specific information either; thus, they fail to exploit any useful information about the data memory hierarchy of embedded systems. In most of the cases, the DMMs of general purpose systems assume one single flat data memory hierarchy. The result is that while these general designs serve a wide range of software application demands equally good, they fail to optimally satisfy the special needs of specific multimedia and network applications mapped on specific data memory hierarchies of an embedded system. As a result, the performance of all those DMMs is very low with respect to custom solutions targeting embedded multimedia systems, and the memory footprint, memory accesses and especially the energy consumption figures are unacceptably high.

Additionally, existing general implementations of DDTs and one-size-fits-all dynamic memory manager solutions are suboptimal because they do not provide specialized solutions to the individual memory storage and resource needs of the software applications, the available memory resources of the underlying data memory hierarchy are not optimally used. Finally, more customizable DDTs and DMM designs lack of a systematic methodology, linked with automation tools, which can enable fast and efficient optimizations targeting embedded system relevant metrics (e.g., energy consumption or memory footprint). In the following, we summarize the set of problems that this book addresses, the solutions that we introduce and the result obtained:

1. **Problem:** Increased design complexity of DDT and dynamic memory manager solutions due to the interdependencies between design decisions.
   Solution: Split the more complex, monolithic design problem into smaller sub-problems of design choices, classify them and analyze their interdependencies.
   Result: The embedded system designer can design very customized DDT and dynamic memory manager solutions according to the particular design constraints of the target system.
   It is important to note that both DDT and dynamic memory manager design are very complicated tasks combining policies of storing and managing data, mechanisms to access the stored data and architectural design decisions. Thus, it would be impossible to calculate the impact of all the possible design decisions on the cost factors and metrics of an embedded system. As a result, we split the design issue in smaller design choices, which have less complexity and their impact can be studied individually. Then, we succeed in dealing with the complex design of DDT and dynamic memory manager solutions by calculating the interdepen-

dencies of the smaller design choices and thus their combination in complete designs of each case. Finally, by using multiple levels of abstractions we manage to customize the DDT and dynamic memory manager designs at various levels of granularity, thus achieving a reasonable trade-off between design-time needed to explore the design space and the efficiency of the optimizations.

2. **Problem:** Constraint of available hardware resources due to cost, physical size and performance limitations.

   Solution: Extraction of key cost functions to abstract underlying data memory hierarchies in the optimization phase of DDTs. Optimization of DMM implementations via customization according to specific software applications and data memory hierarchies.

   Result: Reduced cost of the embedded system and application design within the desired specifications.

   The solution to the second problem is built on top of the solution of the first problem. Thus, it is done after enabling customization of DDTs and DMM design, and then we can tune each of these two layers according to the specific storage and access needs of each individual software application and according to the available memory resources of each data memory hierarchy layer that meets the specifications of the embedded system. These customizations lead to optimizations targeting individual metrics and cost factors or the achievement of trade-offs amongst them.

3. **Problem:** Short time-to-market of the embedded system.

   Solution: Proposed prototype tool framework for automatic exploration of the optimal combination of DDTs and DMM design choices and implementation using a library of modules.

   Result: Exploration of trade-offs and customized DDTs and DMM designs with minimal design-time overhead.

   Fine-tuning the dynamic memory management subsystem at its two different design levels (i.e., DDTs and DMM) according to the characteristic needs of each software application can be very complex and time consuming for the embedded system designer. Hence, we extract the relevant characteristics automatically with the use of our profiling tools and then provide it as input to the exploration tools which explore the different desired combination of DDTs and DMM design decisions automatically. To this end, the design decisions are implemented as modules in a C++ library and defined a very easy-to-use plug-and-play library to create multi-layer DDTs that can be automatically evaluated to find Pareto optimal solutions in a multi-objective exploration (i.e., memory accesses, energy consumption and memory footprint). Similarly, the design decisions to build DMM solutions are defined in a C++ library of layers that can be combined in a single customized DMM design, which is simulated and evaluated versus the desired optimizations, according to the design constraints of each targeted multimedia and network system.

1.5 General Design Flow for Embedded Systems Conception

The proposed DMM design methodology is part of an initial general meta-design flow targeting the complete stack of embedded systems design. A global view of this global meta flow is provided in [34]. In this unified meta flow, we can identify a number of abstraction levels, which each group their respective (sub)steps. The major steps in the system design trajectory are related to the algorithm design, the task-level concurrency, the data parallelisation into regular arrays, and the instruction-level concurrency.

In Fig. 1.10 we show how this book, solving all the aspects related to the dynamic memory management subsystem, is embedded in this unified flow. The earliest stage in the design flow concentrates on generating a correct functional specification of the design. In order to have a complete and formal description of the system, it is necessary to describe the functionality in a specification language, which can be (but need not be) executable. For real-time systems, however, it is not only important to generate the correct results, but also to generate them on the right time instances.

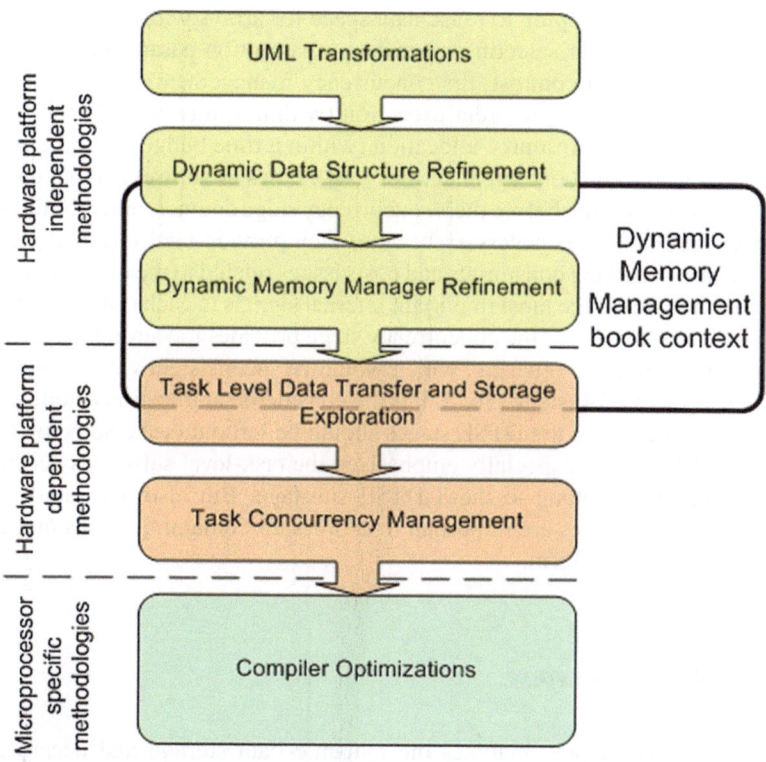

Fig. 1.10 Dynamic memory management in the general design flow context for embedded systems (seen as part of the unified meta-flow [34])

Therefore, the specification model must be able to deal with time and *temporal correctness* as well. An example of such a well-defined language would be UML but also other similar high-abstraction layer languages can be used.

Afterwards, and still belonging to the algorithm design abstraction, we have to refine the dynamic data structures and manage their (de-)allocation. These substages constitute the focus of this book.

Next we have the respective concurrency levels, and at each of those, both the data transfer and storage issues (in the *data transfer and storage exploration* (DTSE) substages [37]) and the concurrency management issues have to be addressed. The DTSE substages are all related to the handling of complex data types such as indexed signals (arrays) or dynamically created graphs (trees or lists). It is assumed though that the DMMR substage has already been applied for the dynamically allocated data structures. The DTSE substages involve global data analysis (for dependencies, accesses and memory size estimates), global data- and control-flow transformations (to remove redundancy and data-flow bottle-necks respectively to increase access regularity and locality), data reuse decisions (related to the exploitation of memory hierarchy), storage cycle budget distribution (to meet timing constraints), memory allocation and assignment (for background memories) and data layout organisation issues (e.g. in-place mapping to reuse data space for arrays with a limited, partially overlapping life-time; or selecting other data organisation parameters in the cache and main memory). In contrast, the concurrency management steps are related to concurrency extraction, concurrent execution (within a time budget), the synchronisation (ordering), the resource allocation (within a time budget) and the interface refinement between the concurrent entities [34]). It has to be stressed that the DTSE stages are always ordered before their concurrency stage counterpart at the three main abstraction levels. This is necessary because they provide vital information on the communication ordering constraints and costs issues related to the background memory storage/transfer of the most important internal signals in each task. Reversing the order would also mean that the concurrency stage provides too strong restrictions on the data communication ordering, which would be inconsistent with the reason for having a separate DTSE stage [36, 37]. If the considered application (module) is not data-dominated enough, the DTSE stages should be left out (see also below).

In Fig. 1.10 we have especially emphasized the task-level substages because our DMA related steps belong to the T-DTSE substage. But also the data-level and operation-level substages are important to address the remaining issues in the mapping process to the platform.

1.6 Chapters Overview

- *Chapter 2* This chapter analyzes the dynamic data storage and access behavior of multimedia and network applications. The behavior is assessed from the dynamic data structure perspective and is extracted with the use of profiling tools.

Moreover, alternative designs for improvement and trade-offs regarding dynamic data structures and memory management techniques are introduced.

- *Chapter* 3 This chapter presents the suitable profiling and application behavior extraction methods for dynamic applications. Among these methods, the use of metadata and profiling templates are proposed as main techniques for raw data behavior extraction. Finally, a complete integrated example where all the different phases of the proposed methodology for dynamic memory optimization of multimedia and network applications is presented.
- *Chapter* 4 This chapter elaborates on the dynamic data type refinement (DDTR) phase of the proposed methodology in the context of multimedia and network applications. This chapter includes the application of the proposed dynamic data structure transformations to three complete applications, Tetris game, 3D-image reconstruction and network scheduling.
- *Chapter* 5 This chapter develops the intermediate variable removal (IVR) phase of the proposed methodology. This chapter analyzes the capabilities to formalize the IVR problem and automate the removal of intermediate variables in the context of very frequently used dynamic data structures in multimedia and network applications, such as, sequences. Finally, it elaborates on the steps for this automation process by demonstrating its application to several examples of very complex dynamic multimedia and network applications.
- *Chapter* 6 This chapter analyzes the dynamic memory manager refinement (DMMR) phase of the presented methodology. More specifically, the complete problem of designing a dynamic memory manager is split into smaller design subproblems and each one is analyzed independently. A search space of design solutions is presented for these subproblems and the effect of specific design choices on the metrics and cost factors is assessed. A highly customized dynamic memory manager can be constructed with the combination of the presented design choices according to specific characteristics of the target software applications and the underlying data memory hierarchy. Then, a set of parameters is presented for some of the design choices, which can further fine-tune the customization of the dynamic memory manager. Next, this chapter describes the interdependencies between the different design choices. These interdependencies are the natural byproduct of splitting the bigger design problem into smaller design subproblems. We explore the constraints that can be passed from one design choice to another and thus enable the faster exploration of the whole dynamic memory manager design search space. Finally, the propagation of constraints is studied according to the metric or cost factor that is pursued for optimization by the exploration. For multiple metrics or cost factors Pareto-optimal solutions are proposed.
- *Chapter* 7 This chapter introduces a flexible methodology to perform the placement of DDTs in the memory. It exploits all the elements of a heterogeneous memory organisation. This methodology identifies the most relevant characteristics of each DDT and groups them accordingly while reducing the memory footprint. The groups of DDTs are transformed into memory pools which are then placed in one or more memory resources. The application of this methodology on a wireless network system case study leads to large average improvements.

Chapter 2
Analysis and Characterization of Dynamic Multimedia Applications

As discussed in Chap. 1, in nomadic embedded systems an increasing amount of applications (e.g., 3D games, video-players) coming from the general-purpose domain need to be mapped onto a cheap and compact device. However, embedded systems struggle to execute these complex applications because they come from desktop systems, holding very different restrictions regarding memory use features, and more concretely not concerned with the efficient use of the dynamic memory. Today, a desktop computer typically includes at least 2–8 GB of RAM memory, as opposed to the 256–1024 MB range present in low-power respectively high-end nomadic embedded systems. Therefore, one of the main steps during the porting process of multimedia applications (that were initially developed on a PC) onto embedded multimedia systems, involves the optimization of the dynamic memory subsystem.

The rest of this chapter is organized as follows. In Sect. 2.1, the multimedia application domain is analyzed including the typical features this type of application possesses, motivating the need for dynamic data optimizations. Then, in Sect. 2.2, dynamic data and its specific properties are presented. Next, in Sect. 2.3, the method flow to tackle the different optimization steps is explained in more detail, covering both the steps and the reasons for the selected ordering of these steps. Finally, in Sect. 2.4 some conclusions are discussed.

2.1 Characteristics of Multimedia Applications

The optimizations presented in this book exploit the characteristics of modern multimedia and communication embedded applications. Where traditional software would only have global or stack data, modern day applications require the use of heap data (or *dynamic data*). In this section, we describe how dynamic data is used in modern multimedia applications, and how the DDTs are used to manage this data. The set of analyzed multimedia and communication applications includes 3D image reconstruction applications [45, 138], video rendering applications as the MPEG-4 *Visual*

© Springer International Publishing Switzerland 2015

D. Atienza Alonso et al., *Dynamic Memory Management for Embedded Systems*,
DOI 10.1007/978-3-319-10572-7_2

Texture Coder (*VTC*) [59, 124], 3D games [71, 122], the URL-based context switching application [116], IPv4 routing algorithms [91] and firewall applications [116].

Due to the fact that these applications deal with data that is not quantifiable at compile-time, they require heap data (or dynamic data) to store the application-required information. For instance, in 3D image reconstruction, studied in the Sect. 2.1.1, two images are compared to find points that match. Since the amount of points that may match is not known at compile-time, a dynamic data-structure is required for this.

2.1.1 Example: 3D Image Reconstruction System

The previously outlined characteristics are illustrated using a modern 3D image reconstruction application [160]. In particular, we focus on one of the internal algorithms that works like 3D perception in living beings, where the relative displacement between several 2D projections is used to reconstruct the 3rd dimension [138]. This software module heavily uses dynamic memory and is one of the basic building blocks in many current 3D vision algorithms: *feature selection and matching*, which involve multiple sequential accesses and input data filtering operations. The algorithm studied has been extracted from the original code of the 3D image reconstruction system (see [160] for the full code of the algorithm with 1.75 million lines of high level C++), and creates the mathematical abstraction from the related frames that is used in the global algorithm. The algorithm selects and matches features (corners) on different subsequent frames and the relative offsets of these features define their spatial location (see Fig. 2.1). The operations performed on the images are particularly memory intensive, e.g., each image with a resolution of 640×480 uses about 2.5 MB, and the accesses of the algorithm to the images are randomized. Thus, classic image access

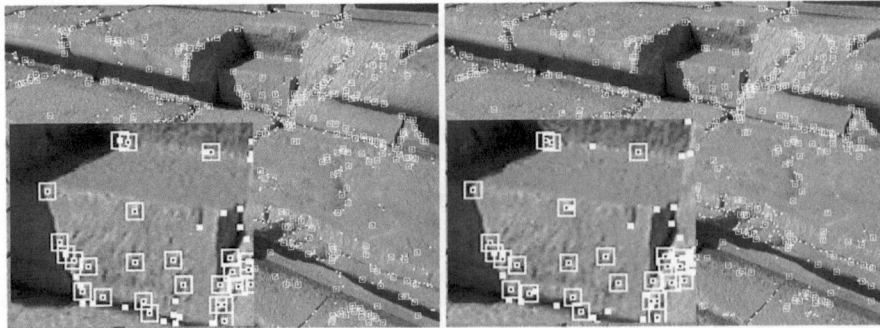

Fig. 2.1 Initialization of the matching of corners on two images of the steps of amphitheater (archaeological site): based on neighborhood search. Already most matches seem to be correct (partially due to the minor difference between the images, which can be seen at the *right hand bottom corner*). Part of the center is enlarged

optimizations such as row-dominated accesses versus column-wise accesses are not applicable.

The number of generated candidate matches is highly dependent on a number of factors. First, the *image properties* affect the generation of the matching candidates. Images with highly irregular or repetitive areas will generate a large number of localized candidates, while a low number will be detected in other parts of the image. Second, the *corner detection parameters* have a profound influence on the results (and, consequently, on the input to the subsequent matching algorithm) because they affect the sensitivity of the algorithm used to identify the interesting features in the images/frames [72]. Finally, the *corner matching parameters* that determine the matching phase have a profound influence and are changed at run-time. For example, the acceptance and rejection criterion changes over time as more 3D information is retrieved from the scene.

Taking all the previous factors into account, the possible combinations of parameters in the system make an accurate estimation of the memory cost, memory accesses and energy dissipation at compile time very hard or even impossible. This difficult to analyze and hence *unpredictable* memory behavior can be observed in many state-of-the-art 3D vision algorithms and makes them very difficult to optimize with traditional compile-time optimization methods for embedded systems. Nevertheless, since this metric 3D-reconstruction from video algorithm can perform the reconstruction of 3D scenes from images and requires no other information apart from multiple frames, it is especially useful for situations where extensive 3D setup with sensitive equipment is difficult, e.g., crowded streets or remote locations, or impossible as when the scene is no longer available [45, 60]. Hence, it is extensively used for quick on-site visualization, and speeding up the application to be able to process more frames for a more detailed reconstruction is a very important problem, which demands extensive code transformations and optimizations of the dynamic memory subsystem. Also, energy consumption is paramount for hand-held visualization devices and needs to be optimized as well. In the following paragraphs the internal DDTs of this case-study are explained and their respective dynamic allocation behaviors are carefully analyzed.

The algorithm internally uses several DDTs whose sizes cannot be determined until the system is running because they depend on factors (e.g., textures in the images) determined outside the algorithm (and uncertain at compile-time). Furthermore, due to the image-dependency related data, the initial DDT implementations for the variables do not fit in the internal memory of current embedded processors. These DDTs are the following:

- `ImageMatches` (`ImageMatches`) is the list of pairs where one point in the first image, matches another one on the second image based on a neighborhood test [138].
- `CandidateMatches` (`CandidateMatches`) is the list of candidates that must go through a normalized cross-correlation evaluation of a window around the points [160].

- `MultImageMatches` (`MultiMatches`) is the list of pairs that pass the aforementioned evaluation criterion. Still one point from the first image can be listed in multiple candidate pairs for a later best match selection.
- `BestMatches` (`BestMatches`) is the subset of pairs where only the best match for a point (according to the test already mentioned) is retained.

These DDTs were originally implemented using double linked lists and exhibit an unpredictable memory size, typical in many state-of-the-art 3D vision systems [160]. That is due to their use of some sort of *dynamic candidate selection* by traversing the input data in a sequential way, followed by *a criterion evaluation*. Figure 2.2 shows this behavior in the generation order of the DDTs. First, `ImageMatches` is generated after a neighborhood test is applied to pairs of corners detected in two images. Then, `CandidateMatches` is created using the information acquired from previous images. Finally, `MultiMatches` and `BestMatches` are generated with the pairs that pass a normalized cross correlation evaluation [138].

Although the global interaction between these DDTs is defined by the algorithm (see Fig. 2.2), it is not clear how each DDT affects the final system figures (e.g., memory footprint) until the application is profiled. Therefore, the method presented in Sect. 2.3 has been used to explore the behavior of the different DDTs. After inserting the profiling library and running the tools presented in Chap. 3, profiling information is obtained. Then, a memory use graph is automatically generated and memory accesses, memory footprint and energy dissipation figures are calculated (see Table 2.1).

When looking at the memory behavior of the application in Fig. 2.3, it is shown that only very few allocation sizes are very popular. This is of course a factor of the DDTs that were originally used in the application. The block sizes used in the DMM, that affect its design, are in turn affected by the DDTs basic memory requests. Because of this property, DDTs should always be decided upfront and they should take into account the restrictions on the available memory organisation. On an existing platform, which is the most common case now, that memory organisation is namely predefined. Hence it forms a constraint that we should incorporate in each of the

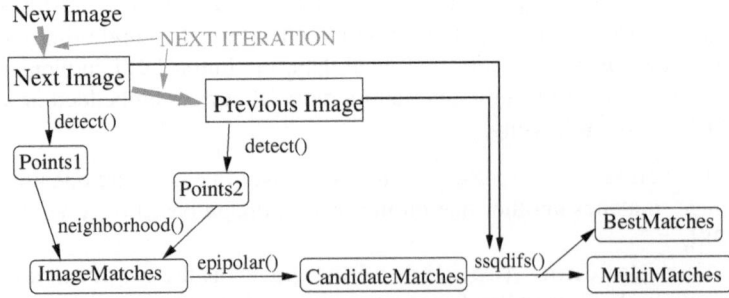

Fig. 2.2 Execution flow of the DDTs in the 3D image reconstruction application

Table 2.1 Original DDT in the 3D image reconstruction application

Variable	DDT accesses	Memory footprint (B)	Energy 0.13 μm tech. (μJ)
ImageMatches	1.20×10^6	5.14×10^2	0.18×10^3
CandidateMatches	8.44×10^5	2.75×10^5	3.03×10^3
CMCopyStatic	6.24×10^4	1.08×10^5	4.48×10^4
MultiMatches	1.84×10^4	3.62×10^2	0.02×10^1
BestMatches	1.66×10^4	3.07×10^2	0.02×10^1
Total	2.14×10^6	3.86×10^5	4.80×10^4

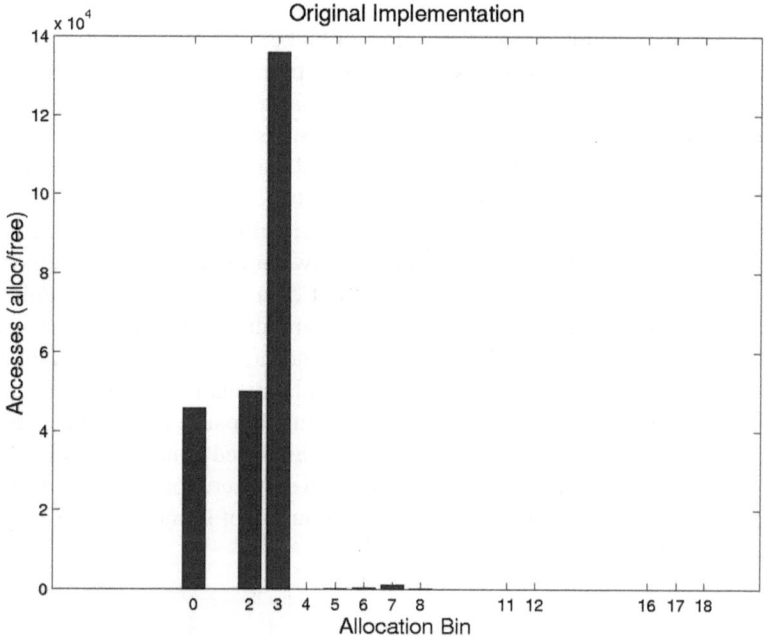

Fig. 2.3 Study of block-size allocation bins of the Kingsley DM manager requested by the original 3D reconstruction application

stages and steps. The DMM basic sizes will then be later decided during the DMMR step, but due to the early DDT decisions that step then has a better starting point.

In order to account for the varying memory use during the program run, normalized memory is used to get an estimation of the overall contribution of each DDT to the energy dissipation shown in Table 2.1. This attributes more accurate contributions to energy cost estimates and avoids that, for instance, DDTs with very short and high memory usage distort and hide the memory contribution from other DDTs. For the energy estimations, the embedded SRAMs memory model described in [84] is used, considering a 0.13 μ technology node for the memory hierarchy implementation. This model depends on memory footprint factors (e.g., size, internal structure or leaks) and factors originated by memory accesses (e.g., number of accesses or technology node used). Note that any other energy model can be used just by replacing that module in the tools.

The analysis of the profiling information (Table 2.1) shows how the DDTs affect the dynamic memory subsystem. The energy numbers are calculated based on a memory that is large enough to encompass only that single DDT. The DDTs and their different dynamic behaviors can be derived from the internal subdivision of the global algorithm in different sub-algorithms previously described, which include different traversal and filtering phases. First, `CandidateMatches` is the largest DDT in this system. Secondly, `ImageMatches` has frequent accesses. Finally, `CMCopyStatic` is a dynamic array implementation, which has a much faster access to specific stored data, that only keeps a copy of the content of `CandidateMatches`) consumes an important part of the energy used by the system. The reason it has such a high energy cost is that certain accesses, namely adding new elements to it incur an O(N) memory access cost. As this analysis shown, the DDTs in multimedia applications interact as local variables (with very limited lifetime) that store the intermediate data generated between the different sub-algorithms that conform the global algorithm (e.g., `ImageMatches` or `CMCopyStatic`), or as variables that store the filtered input data between the different sub-algorithms. These are used in the end to provide the output data of the current module to the next software modules of the overall applications, due to modularity in software engineering, as it is the role of `BestMatches` and `MultiMatches`. A final characteristic of multimedia and communication applications is the repeating cycle of their dynamic allocation and de-allocation patterns. In particular, in the considered 3D image reconstruction module, this previously analyzed dynamic behavior is similar for each pair of consecutive frames, as in each new iteration of the algorithm, the oldest frame is replaced by the next one in the input set of frames and the algorithms are applied in the same way.

2.1.2 Potential for Optimizations

In this book we look at an important subset of the multimedia application domain, namely image processing (image analysis, image interpretation and image reconstruction), video processing (namely multiview video reconstruction) as well as 3D reconstruction and rendering. In general, the more standard image processing (filtering and coding) part of the applications manipulate quite static stack data [111]. However, more advanced computer vision and image understanding algorithms contain a significant part of dynamic data that have to be managed on the heap [45, 99]. The dynamic data types used for this are either based on libraries like the Standard Template Library (STL) [145, 157] or on own code. However, in both cases, the dominant DDTs are fully deterministic (sequences) in their access behavior and not semi-random data-dependent. Hence, we can conclude from this that the majority of these DDTs would be suited for handling by our approach.

Additionally, many image and video processing algorithms operate on data structures which are more complicated than arrays and which are dynamically constructed (e.g., with data dependent record sizes). For instance segmentation algorithms operate on data structures which can be organized as linked lists. Nevertheless, the most efficient algorithms aim to access such data in a regular fashion and avoid random access where possible. For instance, Philips has developed such algorithms, which iterate over variable-length line segments, but in a line-scan fashion. The main application domain is 3D image segmentation.

We also look in more detail at the communication protocol applications for nomadic embedded systems. More specifically, in [18], it is shown how communication protocol applications contain data structures that are dynamic in nature. More specifically, packet scheduling algorithms have to deal with a variety of queues of packets, and these are by definition dynamic in nature, as it is not known in advance how many packets will fall into each scheduling queue.

For these two specific domains, the range of possible applications is in theory initially very broad. However, for most of the embedded systems the number and types of multimedia and communication protocol applications to be included in the final design (at least to a large extent) are known at design-time. Thus, it is feasible to analyze the types of dynamically-allocated objects that exist in each of them (e.g., points, triangles, 3D faces, acknowledgment or frequent packets sizes, etc.) and to design the most convenient DDT implementation for each variable, as well as the most convenient DMM for the application in question.

Furthermore, our experience has shown that, in each application, the range of sizes used by the dynamic elements is very limited and it can be determined at design time by, first, a static analysis of the source code of the application and, second, a profiling analysis of the application under exploration with a reduced number of input data sets (i.e., in general no more than 10 variations). Note that the size of the dynamic elements is known a priori, but not the actual number to allocate, which can vary significantly from one possible representative input to another; thus, the use of dynamically allocated memory is justified and extensively used in these application domains. For example, in order to analyze an MPEG-4 video player, different system configurations, such as screen resolution or visualization size, need to be studied according to the target embedded device: a smartphone, a Portable Digital Assistant (PDA), a portable video-game station, etc. This set of configurations needs to be explored for a representative set of input frames (e.g., with a different number of rendered objects and textures, etc.), while taking into account the probability distribution of the different types of inputs for the final working conditions of the target embedded device, as these features will influence the final dynamic data structure to be used for the final application.

In addition, it is necessary to recognize the dominant dynamically-allocated elements of each application. In fact, each application can contain a large number of allocated elements, but in most of the multimedia and communication protocol applications few variables (i.e., between 5 and 20) tend to account for a very large percentage of the overall memory accesses and/or memory footprint used for dynamic memory storage: in general between 50 and 70 %.

2.2 Dynamic Data Handling

As shown in Sect. 2.1, in modern multimedia and communication protocol applications, data is stored in entities called *Dynamic Data Types* (*DDTs*) or simply containers, like vectors, lists or trees, which can adapt dynamically to the amount of memory used by each application [173]. These containers are realized at the software architecture level [99] and are responsible for keeping and organizing the data in the memory and also servicing the application's requests at run-time. These services require to include abstract data type operators for storing, retrieving or altering any data value, which is not tied to a specific container implementation and which shares a common interface (as in STL [145]).

Dynamic data is commonly referred to as *heap data*, while static data usually either resides in the global data segment or on the *stack*. The difference between dynamic data types and dynamic data structures is that the former provide an interface with a given set of operations, while the latter provide a specific implementation that adheres to the interface. For example, STL provides *sequences* as dynamic data types. Vectors and lists are then two specific data structures with different complexity trade-offs that implement the sequence data type.

To understand the difference between dynamic data types and traditional static data types, it is necessary to understand how individual elements stored in these data types are mapped to the actual memory hierarchy. The mapping of an element to its actual place in memory for data types, both static as well as dynamic, consists of two layers. The first layer, from now on referred to as *element mapping*, maps the location of an element in a specific data structure to a location in the memory block(s) that are employed by the data structure. The second layer, from now on referred to as *block mapping*, maps the data structure's memory block(s) to the memory pools that can represent the physical memory hierarchy. The presence of a memory management unit (MMU) adds a third, hardware, layer, that maps virtual memory addresses to physical memory addresses.

The difference between dynamic and static data types is how these mappings are realized. The differences are first detailed at element mapping layer. Then, the difference are detailed at the block mapping layer.

At the element mapping layer, the mapping is fixed for static data types. The most typical static data structure, the array, demonstrates this quite clearly. The memory-offset of an element in the memory-block used by the array is a linear mapping of its index (namely by multiplying by the size of the element). This mapping is static and defined by the platform's application binary interface (ABI) [30].

For dynamic data types, on the other hand, the mapping from element to location in the memory block is defined by the implementation of the specific data structure. Additionally, through the operations provided by dynamic data types, this mapping can change at run-time. For instance, a vector, which is most similar to an array, maps elements to the memory block it uses in a similar fashion to an array. However, a vector allows for the insertion of elements at any index. This insertion affects all elements after the element inserted at, and changes their location. The fact that dynamic data types allow for the insertion, as well as removal of elements, has implications for the block mapping layer. It is necessary to replace smaller memory blocks for bigger memory blocks as more elements are added to a dynamic data structure. As such, for dynamic data types, the block mapping layer is necessarily dynamic.

For static data types, it is still possible to have a dynamic block mapping layer, in the case that they are allocated at run-time. However, because there is no dynamism in the element mapping, this allocation is necessarily not required as often, as it is only required to create entirely new data structures, not to deal with the dynamism at the element layer mapping.

The actual implementation of dynamic data structures only encodes the element mapping layer. The dynamism at the block mapping layer is handled by the *Dynamic Memory Manager* (*DMM*) which takes care of not only servicing memory allocation requests but also ensures this is done in an optimal manner.

With the combination of both the index-layer mapping as well as the block-layer mapping, elements of DDTs are thus stored into memory pools. After this mapping, it is then possible to map these memory pools, along with the pool dedicated to stack data and global data onto the actual real memory hierarchy. This step combines the dynamic and static data types and is complimentary to and compatible with the index-layer mapping and block-layer mapping. Therefore, dynamic data must be tackled first before it can be combined in terms of memory pools with the remaining static stack pool.

As shown, two components are required to enable the design of applications that use dynamic data. First of all, the implementation of the operators of a DDT can have significant impacts on its access and storage patterns. Secondly, the implementation of the DMM can have an impact on where these accesses and storage reside in the memory hierarchy. These two are detailed further below.

2.2.1 Dynamic Data Structure Optimization Opportunities

The design of DDTs is dictated by a specific interface composed of a set of operators. Existing and widely-used libraries have a good set of operators that provide the required flexibility without posing too many constraints on the implementation of the DDTs (e.g., STL [145] and Java Collections). Before we look at these operators, it is important to realize that also a variety of different types of DDTs exists. And the choice of which DDT is ideal, is usually based on the application in question. As such, a wide set of libraries exist that give different classes of DDTs. Some libraries, like STL [145] are more standardized and come with most C++ compilers. Other libraries are more fragmented, and are thus ad-hoc standards or built for specific purposes. These two categories of libraries are studied in more detail.

2.2.1.1 STL

The classification, as defined by STL, consists of the following top-level concepts. This hierarchy, along with the implementations is also presented in Fig. 2.4. The darker elements are the different classes of DDTs, while the lighter elements specify specific implementations provided by the STL library.

- **Container** A Container is an object that stores other objects (its elements), and that has methods for accessing its elements. In particular, every type that is a model of Container has an associated *iterator* type that can be used to iterate through the Container's elements. [145].
- **Sequences** A variable-sized container whose elements are arranged in a strict linear order. Supports insertion and removal of elements.
- **Associative Containers** A variable-sized container that supports efficient retrieval of elements based on keys. Supports insertion and removal of elements, but differs from sequences that it does not provide a mechanism for inserting an element at a specific position.

Besides these generic categories, STL also provides some specific data-structures, mostly to provide functionality on top of the above-described set of containers, or to address the specific needs of handling strings. Due to the fact that these are either based on top of the existing containers, or address a very specific need (string handling) that is less relevant to the multimedia domain, they are not further treated in this book.

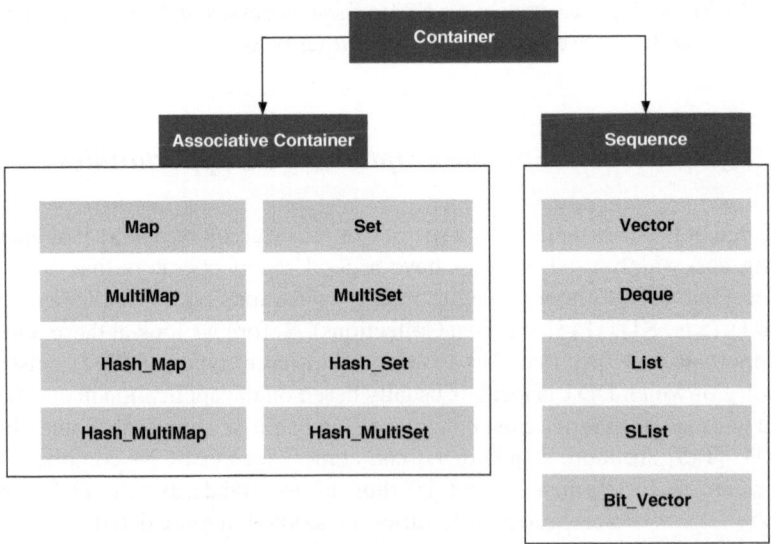

Fig. 2.4 Hierarchy of the different DDTs as classified by STL (in *darker color*), along with the implementations (in *lighter color*) that come with STL

- **Ropes** These are designed to be scalable string implementations and are specifically designed for operations that involve a string as a whole.
- **Bitset** Very similar to a vector of booleans, it contains a collection of bits and provides constant-time access to each bit. Unlike a vector, however, the size of a bitset cannot be changed. As such, this makes it much more akin to an array. Bitsets are not covered in this book as they are not defined to be an STL container, since they lack iterators for traversal.
- **Container Adapters** These are not technically containers, but instead adapters that can be placed on top of other containers to restrict the interface.

Relevant is the fact that accesses to the containers are achieved through iterators. Where static data structures like arrays employ the use of pointers for accessing and navigating through the elements, this is only possible because the memory locations of subsequent elements are contiguous. Iterators are a generalization of pointers. Iterators are central to generic programming because they are an interface between containers and algorithms: algorithms typically take iterators as arguments, so a container need only provide a way to access its elements using iterators [145]. The iterators are categorized by STL, as shown in Fig. 2.5.

- **Trivial Iterator** An object that may be dereferenced to refer to some other object. The simplest example in the case of static data types is a simple pointer into an array.
- **Input Iterator** An iterator that may be dereferenced to refer to some object, and that may be incremented to obtain the next iterator in a sequence.
- **Output Iterator** An iterator that provides a mechanism for storing (but not necessarily accessing) a sequence of values.
- **Forward Iterator** An iterator that corresponds to the usual intuitive notion of a linear sequence of values.

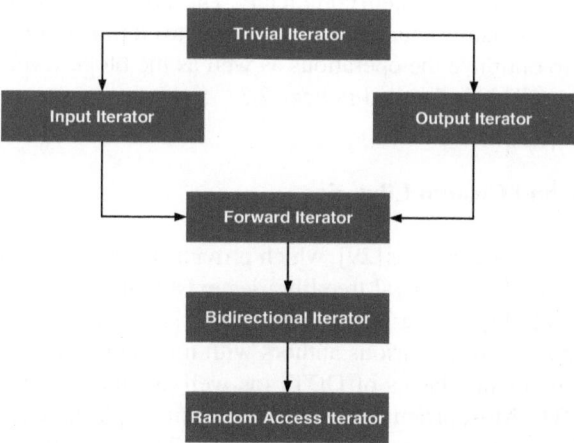

Fig. 2.5 Hierarchy of the different iterator classes used to traverse and access DDTs, as specified by STL

- **Bidirectional Iterator** An iterator that can be both incremented and decremented.
- **Random Access Iterator** An iterator that provides both increment and decrement and that also provides constant-time methods for moving forward and backward in arbitrary-sized steps. The best example of this is a pointer into an array.

The final way that these DDTs can be classified is by how they are used in the application. Here, we discern different ways a DDT can be used in an application. The usage-patterns are based on how the iterators to the DDT are used. While most operations defined on DDTs do require an iterator, some operations do not. For sequences, we consider the operation `append` as based on a forward iterator, as the location where this adds an element is well-defined. On the other hand, assignment at an index for a sequence or assignment of a value to a key in an associative container is considered as based on a random iterator. As such it is possible to define two different usage patterns:

- Static number of iterators accessing the DDT. This is the case when a DDT is used as an intermediate variable and is built up in a set of phases. Note that the actual number at run-time must not be static, but conditionally static, in that there is a well-defined set of loop nests that *could* modify the DDT. Note that this condition is not overly restrictive, as control-flows are mostly static. As such, as long as there is no top-level loop which constantly creates an iterator to access the DDT, the usage pattern falls in this category.
- Dynamic number of iterators accessing the DDT. This is the case when a DDT is used as a central state, and is being modified over and over in a top-level loop. An example of this is a histogram that is being calculated over an entire input, and thus a random access iterator is used for each value of the input to modify the DDT.

Where DDTs that function as central state can be optimized to have efficient operations, for DDTs that function as intermediate variables, it is possible, given the right optimizations, to remove them completely. This is detailed further in Sect. 2.3.3. For all of the above dynamic data types, it is possible to employ information regarding the application to optimize the operations as well as the block-level mapping to the memory hierarchy. This is detailed in Sect. 2.3.

2.2.1.2 Ad-hoc and Custom Libraries

Besides the Boost C++ Libraries [29], which provide a variety of extra functionality missing in C++ and STL, several tree libraries exist that are publicly available for quite a few years [145]. While Boost is an ad-hoc standard, the other libraries are personal contributions from various authors with no standardization or committee. These libraries focus on classes of DDTs (as well as other functionality) that are not present in STL. Most prominently, they focus on graphs and trees, namely data structures that have more structure in their interface. Trees have the concept of parent and child links as well as siblings, where sequences only have sibling elements. Graphs, add even more flexibility to the links.

It is important to note that the discussion here on trees, is for trees where the actual structure of the tree has semantical relevance to the application. Some of the associative containers, for instance *map*, use trees internally, however from an interface perspective they are presented as an associative container mapping keys to values. Trees presented here, however, are used by applications that *rely on a specific structure in the parent-child relationship beyond those employed for pure performance reasons*. As such, these libraries have a lot less flexibility in terms of the internal data structures used for representing these trees. On the other hand, because there is semantics in the physical ordering elements, this information can be exploited for optimizations.

Similar observations here hold as in the last section regarding potential for optimizations. In principle, intermediate versions of tree data representations can be redundant and hence removable. But due to the different interfaces and larger variety in the way these tree libraries are defined, the actual optimization procedure is not worked out in this book. Only the STL sequences are addressed in Sect. 2.3.3. Extending this to sequence type tree DDTs is left as future work.

2.2.2 Dynamic Memory Manager

The task of the DMM subsystem is to supply the application or DDTs of the application with memory blocks where the actual data can be stored. These blocks are requested through *allocation* and are then returned from the application to the DMM subsystem through *de-allocation*. This occurs through operations that are well-defined [171], and these are detailed below. Since the DMM subsystem is only defined as those two operations as well as the expected behavior of those two operations, it is relatively unconstrained in terms of implementation-freedom. Only a few basic guarantees must be provided: that the allocation of a memory block will return a block of at least the requested size; that a memory block that is allocated will not overlap with another allocated memory block; that the memory block is not moved in memory; and preferably that de-allocated memory is reused for later allocations. The two main operations provided by a typical DMM are:

- malloc (called by **new** in C++): An operation that takes as input a required size and returns a memory block that is at least as large as the requested size.
- free (called by **delete** in C++): An operation that frees a previously allocated block, therefore allowing it to be reused for future allocations.

Several non-functional constraints are present that must be taken into account before a DMM can be deemed optimal, or even usable. One of these factors is memory fragmentation. It has been introduced already in Sect. 1.3.1, including the distinction between two types of memory fragmentation namely internal and external. To better illustrate this concept, an example is provided here.

In Fig. 2.6, we have two examples of an allocation/de-allocation pattern occurring, at the top and the bottom. In each pattern, a block of 20 kB is allocated by the application, subsequently de-allocated by the application, and then a 40 kB block is allocated by the application. In both cases, the DMM subsystem returns the first memory block that it can find that fits the required size. In the first case, the DMM subsystem does not split 40 kB free block it finds. This results in a less free memory being available after the allocation than if it were to return a 20 kB block. This is called internal fragmentation and results in unused space that the application will not be able to use as long the 20 kB block is allocated. The DMM subsystem cannot return the remaining 20 kB block as the entire 40 kB block is considered used, and the application cannot use the extra space since it only knows that it allocated a block that can fit 20 kB. In the second case of Fig. 2.6, the DMM subsystem decides to split the 40 kB free block to deal with internal fragmentation. However, this results in the subsequent request for 40 kB to not find any free blocks available. In such a case the DMM subsystem would have to request more memory from the Operating System to be able to comply with the application's request (resulting in a bigger pool of memory blocks). In the meantime, however, the memory occupied by the 20 kB free memory blocks remains unused and thus wasted. This is referred to as external fragmentation. In this simple example, the DMM subsystem could have picked the 20 kB block for the 20 kB request and would have alleviated both fragmentation issues. In real systems, finding the optimally matching free block can become expensive due to the amount of effort of finding that block. The second DMM subsystem could have coalesced the two 20 kB blocks back together into a 40 kB blocks. This also comes with cost trade-offs in real systems. Additionally, coalescing can result in internal fragmentation or external fragmentation at a later point and thus coalescing at a particular point in time may not be globally optimal.

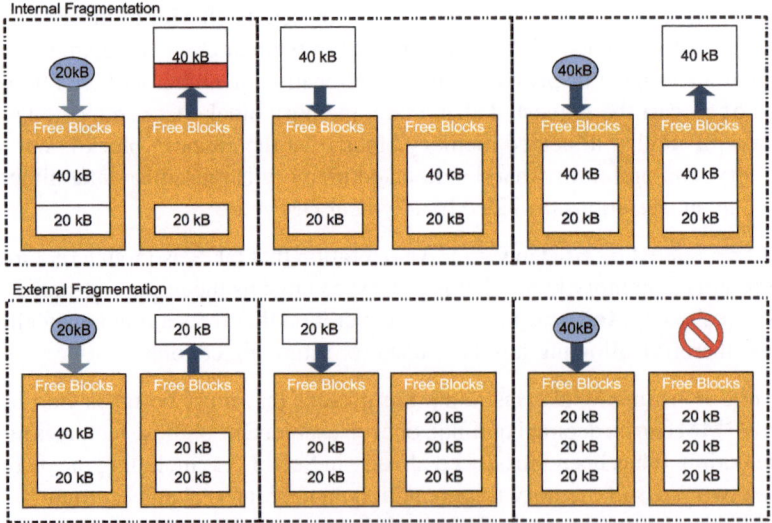

Fig. 2.6 Example of external and internal fragmentation

While most operating systems ship with a standard DMM library [171], these are never optimized for energy consumption. Additionally, they do not take into account the behavior of the application in terms of memory allocations/de-allocations and memory accesses. In the presence of a memory hierarchy such as found in embedded systems, this often results in suboptimal choices for memory allocators [106]. Like the design of DDTs, the design of the DMM subsystem has to take into account several factors before they can be considered useful for a given system:

1. The allocation pattern over time of the application, both directly as well as through the use of the DDTs. It is important that the DMM subsystem is able to efficiently allocate and de-allocate blocks that are frequently allocated/de-allocated. This efficiency is typically measured both in terms of computation as well as the memory accesses that the DMM subsystem performs under the hood for these operations.
2. The memory footprint of the memory allocator in question is not only determined by the dynamic data of the application itself, which obviously is a contributing factor to the dimensioning of the system, but also the memory fragmentation within the dynamic memory manager. If this is not kept in check, the dynamic data can easily consume all the resources of the platform in question, thereby degrading or even completely breaking the expected behavior of the application. As such, memory fragmentation is an important factor for the overall system behavior, that needs to be taken into account.
3. The access pattern of the application to the dynamic data, and hence dynamic memory allocated for it, determines the energy consumption of the application. By placing often accessed blocks onto local and smaller memories it is possible to have a tremendous gain in energy consumption. Thus by taking this access pattern into account, the DMM subsystem can ensure lower energy consumption.

2.3 Proposed Optimization Method

As explained in Sect. 2.2, different optimizations are required to reduce the energy cost incurred by dynamic memory accesses in the application. These optimizations target not only the multimedia applications that need to be improved, but also the libraries as well as the middleware employed by these applications. The method to target such applications is generic, but the steps produce specific results for each multimedia application. By customizing the different aspects like the libraries and the run-time to the multimedia application, it is possible to reduce the energy cost of dynamic memory in these applications significantly. Since the solutions are specific to the application in question, information regarding the application must first be collected though.

The systematic methodology which has been introduced in Sect. 1.4, will now be discussed in more detail. The method that is presented in this book is platform-dependent (i.e., it depends on the instance of the data memory hierarchy organisation used in the platform, but it is processor architecture independent. Since it deals with the energy consumption of memory behavior, it is dependent on the memory sizes that are present in the system. However, the trade-offs of the different optimization steps will not be dependent on the exact energy cost of each access. As such, only the relative energy costs, and thus the relative sizes of the different memory levels (L1, L2, main memory) of the memory hierarchy are of relevance, and not the exact details of the specific architecture in question. While the results and exact final design choices made in the different optimization steps are platform-dependent, the method itself is generic for all platforms, and only the cost-functions and thus the final design choice for a specific design will be platform-dependent, not the applicability of the method itself.

Where in traditional optimization approaches, the information was about static data, which usually is manifest in the source code in terms of array-sizes and loop-boundaries, the introduction of dynamic data removes this source of information. Although production and consumption loops can still be identified, due to the resizability of the DDTs (that serve as storage for this dynamic data) at run-time, it is no longer possible to identify the quantities of data in question. Therefore, the information regarding the behavior of the application in terms of its dynamic memory consumption must be collected at run-time. As such, the method starts with a meta-data collection step before the different optimization steps are applied, resulting in a method flow to tackle dynamic memory [22].

We now look at this method flow, motivating why it is structured and ordered as it is and explaining what the different steps entail. First, the overall method is detailed in Sect. 2.3.1. Then, the step that collects the information required for the different optimization steps is detailed in Sect. 2.3.2. Finally, the different optimization steps are detailed in Sects. 2.3.3, 2.3.4, 2.3.5 and 2.3.6.

2.3.1 Method Overview

In this section we give an overview of the method, explaining the order of the different steps and motivating why this order makes most sense. The explanations of the different steps are then detailed further in the following subsections.

As shown in Fig. 2.7, the first step in the method focuses on the collection of information regarding the behavior of the application. What is not shown in the figure are the high level optimizations that are possible at the object-oriented or higher description levels (e.g., Unified Modeling Language or UML [130]). These need to come first since they can completely alter the design of the application and the existing DDTs. The can be decoupled of the other subsequent steps by introducing high-level estimators [18].

The profiling step information is collected in the form of software metadata, detailing the usage of DDTs and dynamic data in general, both in terms of allocations and de-allocations as well as accesses, and summarized according to a variety of axes.

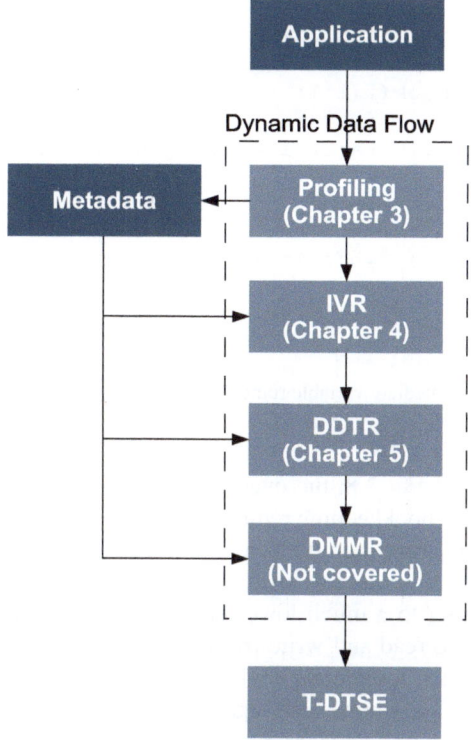

Fig. 2.7 Overview of the proposed methodology

This information is then passed on to the various optimization steps, enabling them. The profiling and analysis step is explained in further detail in Sect. 2.3.2.

After the collection of the information, it is possible to have a holistic optimization with regard to the usage of the DDTs at the application or algorithm level. Typically, DDTs are used in two fashions: As central storage for elements in an algorithm, thus serving as the algorithm's state; or as a buffer to decouple production and consumption phases of an algorithm. In the second case, which is frequently observed in multimedia applications, the use of DDTs is intended to keep algorithms clean and from becoming too complex. In such cases, it is possible to systematically optimize away these DDTs, by a novel optimization step that is termed *Intermediate Variable Removal* (*IVR*). This step enables a trade-off between computation and memory bandwidth. Depending on the platform and the relative costs of memory accesses and computation, this allows for optimizations in term of energy consumption. A small example of IVR is presented here to give the reader some intuition as to what it entails. Further information with regard to IVR is given in Sect. 2.3.3.

In the first code-sample of Fig. 2.8, a DDT is created to which a sequence of elements (generated procedurally) is added. Then, this sequence of elements is consumed in a second loop. Had the two loops been written at once, as shown in the

```
vector<int> a;
for (int i = 0; i < SIZE; i++) {
  if (h(i))
    vector.push_back(f(i));
}
for (vector<int>::iterator i = a.begin(); i != a.end(); ++i) {
  g(*i);
}

for (int i = 0; i < SIZE; i++) {
  if (h(i))
    g(f(i));
}
```

Fig. 2.8 Example of intermediate variable removal

second code-sample of Fig. 2.8, the overhead of first writing all these elements to memory (and the DDT bookkeeping required to make the sequence grow as needed) as well as the reading from memory would not have been necessary. Instead each element could have been written to local memory (or even a register) and then consumed directly, leading to a much lower energy consumption as larger memories require more energy to read and write from. Additionally, the internal DDT bookkeeping would have been completely obviated. IVR focuses on such transformations, ensuring to not break constraints with regard to the ordering of input/output as well as dealing with more complicated production and consumption loops.

Only once it has been decided which DDTs actually remain, it is effective to start optimizing their implementation to a Pareto-optimal design taking into account the platform that the application is being mapped on as well as the resulting behavior of the application in terms of the DDTs. DDT optimizations are not discussed further in this book as they are not part of the core research of this book. Instead I refer the reader to [16, 18]. Again, high-level estimators are used to decouple this step from subsequent steps [18]. Taking the example again in Example 2.8, if the intermediate DDT had not been removed, ideally a list-like DDT would have been used as it leads to less copying when constantly adding elements to the end of it.

Finally, once the implementations of the DDTs are defined, it will be clear what blocks are allocated by these DDTs as well as the application itself. Since the DDT implementations define the internal storage policy on memory blocks, it is required to have the final implementation of the DDTs to see what the allocation behavior of the application looks like. For instance, a simple switch such as moving from a vector style sequence to a linked list style sequence not only fundamentally changes the allocation pattern, but also the types of blocks allocated. In the case of a vector, a contiguous memory block containing the elements will be allocated, resulting in one big allocation (or several as the vector slowly grows and the memory block is expanded and thus reallocated). On the other hand, if a linked list is used, many small blocks will be allocated, each containing a single element and pointers to the previous and next block, with an allocation occurring each time an element is added. Thus, it is clear that the DDT implementation has a big impact on the allocation pattern. This is why the DMM optimisation step comes after the DDT optimisation step. Further information with regard to the DMM optimisation is given in Sect. 2.3.5.

Note that all the above steps deal with dynamic data and how to map this to memory pools. After these steps, it is then possible to combine these abstract memory pools with the global and stack data. This falls outside of the scope of this book, but it shows how the previous method ties into existing work on static data. More information can be found in Sect. 2.3.6.

2.3.2 Profiling and Metadata Collection

To enable the different optimization steps, as already explained in Sect. 2.3, it is necessary to first understand what the dynamic memory behavior of the application is. To this end, since the target application is dynamic, it is necessary to profile it to get an accurate view of the different demands of the application in terms of memory accesses and memory allocations. Additionally, higher abstraction-level profiling information is also captured, such as the use of the different methods provided by a DDT. By using a standard interface (e.g., the sequence interface from STL [145]) it is possible to study the behavior of the application in terms of this interface and then optimize the DDT implementation based on how the DDTs are used. The profiling library that has been developed captures all this information without requiring a drastic change in the application source code [47].

After the raw profiling information is captured, it is analyzed and summarized into metadata that describes the memory behavior of the application at a high level. Typical information here will be how the different DDTs in the application are used, what the typical allocation behavior of the application is, where the most memory accesses occur, etc. Performing this profiling and analysis for a set of representative inputs, it is possible to get a clearer picture of the application's behavior. With this information, it is then feasible to identify intermediate DDTs that can be removed, optimize the implementation of the remaining DDTs and define an optimal memory allocator for the given application and hardware platform. More information regarding profiling and analysis of the application behavior can be found in Chap. 3.

2.3.3 Intermediate Variable Removal

Prior to deciding what the implementations are for the different DDTs in the application, it is first necessary to determine which DDTs are actually required for the algorithm to run. From a design point of view, it is a good idea to decouple production from consumption statements, as it creates a clear separation with a standard interface, namely the intermediate DDTs between these different steps in the application. Some DDTs are required from a theoretical stand point as they can encode the central state of a certain part of the application, where the application cycles in a loop and regularly updates the state. On the other hand, other DDTs only exist in specific phases as temporary buffers between two steps of an algorithm.

From the point of view of the modular programming paradigm and the use of abstract data types [5, 32], which are used in all modern programming languages, the functionality of any program should be decoupled and defined independently from the specific implementation of the data structures (dynamic or static) used to store the application data, as long as all the methods and access functions required by the implemented algorithm are available [82]. According to this software programming paradigm, the proposed method relies on the assumption that the DDTs can be identified in the original source code of the application under exploration, and that they can be replaced without collateral effects that require the modifications of the control flow in the application. In other words, the input source code of the application where the presented method is applied must not include the DDT implementations, and they must have been defined in a separate module or standard library of DDT implementations in C++, such as STL. Moreover, all the algorithm using the DDTs implementations must interact with them only through the standard and common set of access methods [6].

It must be noted that the aforementioned restriction does not impose any real limitation with respect to the applicability of the underlying method, as all the studied applications and most of the applications nowadays, specially designed using the object-oriented paradigm, are implemented following this assumption. Moreover, as the object-oriented programming paradigm is being adopted further as the common design standard to exploit the System-on-Chip paradigm at the software level, this previous assumption will need to be enforced further, due to the necessity to define and use reusable libraries, based on standard interfaces.

The IVR optimization step uses information from the profiling step to determine which DDTs are being used in a producer-consumer like fashion. This is one of the reasons why the profiling step not only profiles memory behavior but also behavior of the application in terms of abstract DDT operations as defined by the DDT interface. More information regarding the IVR optimization step can be found in Chap. 5. Only once it is known which DDTs remain, it is possible to move on to the next step, namely the optimization of DDTs, which is not presented in more detail in this book. More information regarding DDT optimizations can be found in [18].

2.3.4 Dynamic Data Type Refinement

Typically a trade-off exists between DDTs that are mostly used in a random access fashion and DDTs that are mostly accessed in a sequential fashion. It is not possible to design DDTs that support both types of operations in $O(1)$. To give the reader a feeling of this trade-off, Table 2.2 gives a few standard implementations provided by most generic libraries. It should be noted that these trade-offs are only at the granularity of complexity (which translates into both memory accesses and computation). However,

Table 2.2 Typical complexity trade-offs between sequence implementations often found in standard libraries

Implementation	Random access	Append	Prepend	Insertion	Removal
Vector	O(1)	O(1)	O(N)	O(N)	O(N)
List	O(N)	O(1)	O(1)	O(1)	O(1)
Fingertree	O(log N)	O(1)	O(1)	O(log N)	O(log N)

once energy consumption is taken into consideration, memory footprint as well as other factors come into play. These factors are detailed further below.

The implementation of the DDT operations depends on the chosen container and each one of them can favor specific data access and storage patterns. Each application can host a number of different containers according to its particular data access and storage pattern in the algorithm. Choosing an improper container implementation for an abstract data type will have significant negative impact on the dynamic memory subsystem of the embedded system [16, 18]. On the one hand, inefficient data access and storage operations cause performance issues, due to the added computational overhead of the internal DDT mechanisms. On the other hand, each access of the DDTs to the physical memory (where the data is stored) consumes energy. So unnecessary accesses can comprise a very significant portion of the overall power of the system [21, 52]. In fact, energy consumption is one of the most critical limiting factor in the amount of functionality that can be placed in these devices, because portable computers like tablets and notebooks running complex multimedia applications rely on limited battery energy for their operation. As a result, the design of the DDT for a certain nomadic embedded system and multimedia application needs to consider several combined factors:

1. The algorithm implemented in terms of the high level operators that the DDT provides. Since different implementations will have different costs for the variety of operations, this is one of the main factors. While certain implementations may have O(1) access, they will typically require O(N) for insertion, where 'N' is the size of the DDT. A plethora of options exist, but no DDT operation has O(1) for all operations and thus a trade-off exists here between the different operators. This cost is expressed both in terms of computation as well as number of accesses, and thus it heavily affects the overall energy consumption.
2. The storage requirement of the DDT implementations also varies a lot based on the implementation chosen. Certain implementations require extra pointers to enable the management of the underlying memory. Others require no such overhead but do not allow splitting the memory used by the elements into smaller blocks, thereby potentially disabling putting part of the elements onto a smaller (and thus cheaper) memory. These trade-offs have a big influence on both the memory footprint as well as the mapping to the memory hierarchy, and indirectly also to the energy again.

3. The access behavior, or traversals of the elements inside a DDT, specifically consumption of the data, also has a big influence on the design choice of the DDT. As shown, DDTs can be consumed in a variety of ways. This can have a big impact both in terms of complexity in computation and memory accesses required for such traversals, as well as on the caching-behavior of the DDT in question.

2.3.5 Dynamic Memory Manager Optimizations

The final optimization step for dynamic data is the optimization of the dynamic memory allocator. This step should come after the optimization of the implementations of the DDT as these will affect the allocation and access pattern of the application. In the dynamic memory manager refinement (DMMR) step, information about the application in terms of allocation patterns as well as access patterns to the allocated blocks is used from the profiling metadata to determine which memory block types are more heavily used, and which less. With this information, it is possible to determine how to structure the memory allocator in terms of two specific aspects.

First, it helps with the design of the structure of the memory allocator. If many different block sizes are used by the application, then splitting and coalescing is definitely beneficial as otherwise a lot of memory footprint will be lost to fragmentation. Additionally, it helps to determine which memory blocks are often allocated, thereby allowing the decision as to which memory blocks should have their own custom freelists for quick allocation of these blocks.

Second, it helps with the placement of the memory blocks on top of memory pools. Memory pools are typically large blocks of memory that are placed on different elements of the memory hierarchy. The simplest approach is to have a memory pool per memory hierarchy element. Frequently accessed blocks should then be placed onto smaller memory pools that are closer to the CPU, while less frequently accessed blocks should be placed on memory pools further away from the CPU. A good metric is the number of accesses per byte, since it does not make sense to place a frequently accessed block on a smaller memory pool if this block is larger than two smaller pools that combined have more accesses.

To enable this entire optimization step, some exploration is required to find out the exact parameters that give the optimal memory allocator. As such, it is necessary to have a library of combinators that allows one to span the design-space of memory allocators. This set of combinators should be easily composable such that memory allocators of any complexity can be built instead. This book presents such a library of composable blocks for designing memory allocators in Chap. 6.

2.3.6 Task-level Data Transfer and Storage Exploration

After the previous platform-independent steps are done, there is an optimal mapping from the application's dynamic data types to memory pools. Using the memory pools that result from this heap data, along with the memory consumption and accesses due to stack and heap data, it is then possible to map this onto an actual memory hierarchy considering the interaction of different sets of tasks. This final phase, called Task-level Data Transfer and Storage Exploration (T-DTSE), is covered in [36]. However, the part covered in this book related to physical memory allocation and memory assignment (Chap. 7) is complementary to T-DTSE as it enables the application of access ordering, combined with several platform dependent enabling source-code transformations. These steps are possible thanks to the dynamic memory allocation and optimization phases described in the other chapters of this book (Chaps. 2–6).

2.4 Conclusions

In this chapter we have seen how the latest multimedia applications targeting new nomadic embedded systems (e.g., 3D games, video-players) share many features regarding multi-task execution and complex memory management, which makes them ideal candidates for dynamic data optimizations. These complex and dynamic implementations are derived from the initial desktop systems, which have much more memory and capabilities. Therefore, we have analyzed in this chapter the features of dynamic data in this set of new multimedia applications, and proposed a new design flow to optimize and map their dynamic memory management in nomadic embedded systems.

Chapter 3
Profiling and Analysis of Dynamic Applications

As explained in Chap. 2, the problem of optimizing the design of dynamic embedded systems lays on exploiting as much as possible static (design-time) knowledge on the applications, but still leaving space for run-time considerations that allow tackling dynamic variations without resorting to worst case solutions. This requires extensive information about the static and dynamic characteristics of the applications. However, there is not a standard definition or representation of software metadata to typify the characteristics of the dynamic data access behavior of applications because of varying inputs.

Figure 3.1 illustrates how the use of a common information repository (i.e., software metadata) can reduce the overall effort required to apply different optimization techniques on an application. Without metadata, each optimization team has to study the application independently to extract the characteristics that are relevant for their work. Although the information needed may be slightly different for each team, it is also quite possible that common areas exist. Therefore, even if the process of metadata mining may be more complex than the process of extracting the information required by each of the teams in isolation, the accumulated effort can be reduced significantly. Moreover, once the metadata has been extracted, it is available at no cost for any later optimization works, which introduces significant time savings if design teams work iteratively on the same application source code.

In this chapter, a uniform representation of the dynamic data access and allocation behavior of the applications is proposed, defined as *software metadata*. Additionally, this chapter outlines profiling and analysis methods to systematically obtain these metadata. The selected metadata provide a system-level representation of the behavior of dynamic embedded software applications. The contributions of this chapter are:

1. A uniform representation for the application metadata and a method to extract it from dynamic applications.
2. Profiling and analysis techniques that demonstrate a concrete implementation of this method.

© Springer International Publishing Switzerland 2015
D. Atienza Alonso et al., *Dynamic Memory Management for Embedded Systems*,
DOI 10.1007/978-3-319-10572-7_3

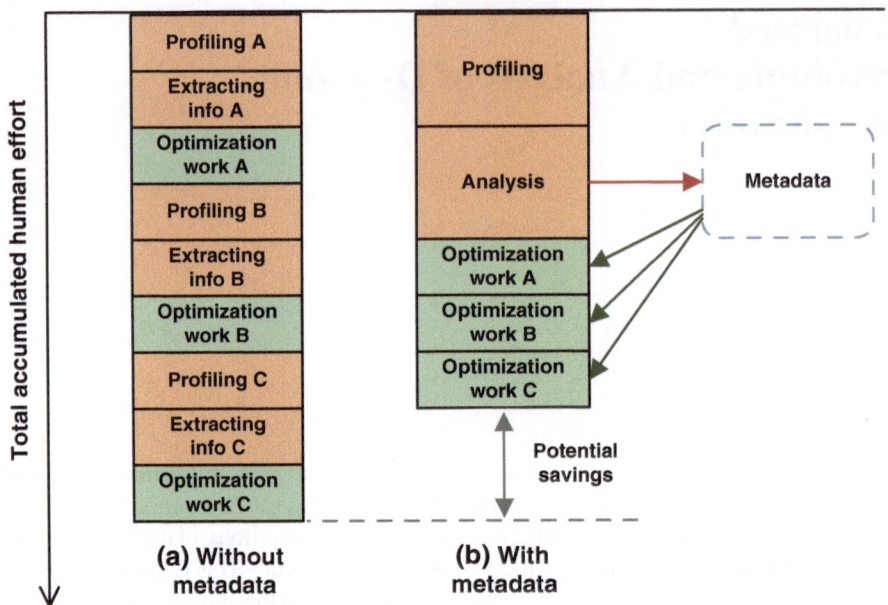

Fig. 3.1 Qualitative comparison of the effort/time invested when using the proposed sharing of application metadata or a traditional design flow without a common information base

3. An example of how these metadata can be used by different system level methods and their relevance to the targeted memory management optimizations.

   The rest of the chapter is organized as follows. First, the metadata representation is introduced in Sect. 3.1. Next, a specific method to obtain the profiling information and the analysis techniques to turn it into the desired metadata are discussed in Sect. 3.2. Then, in Sect. 3.3, one case-study is shown that employs this metadata for various optimizations that concern the dynamic data energy consumption and memory footprint. Finally, an overview of related work is provided in Sect. 3.4, and the conclusions and potential future work in this area are outlined in Sect. 3.5.

3.1 Software Metadata Structure

We propose to classify the information about a software application in three levels of increasing abstraction:

Level zero metadata: Extensive characterization generally obtained through profiling.

Level one metadata: Aggregate representation of the information in the previous level created by the analysis tools. This information is used and updated by the optimization tools during design-time.

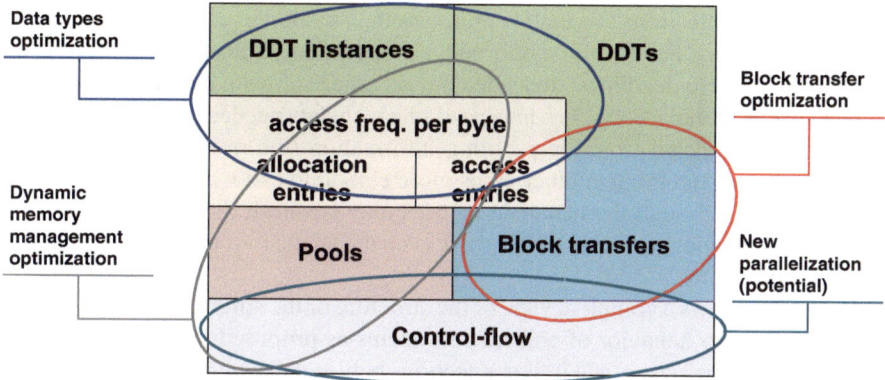

Fig. 3.2 Various optimization tools can use different, but potentially overlapping, subsets of the software application metadata

Level two metadata: Information that is deployed with the final application detailing its resource needs. The run-time manager of the embedded system can use it to adapt the performance of the system to the needs of the applications currently running. This information can be completed with the variations of available resources during time (e.g., battery capacity, new modules that are plugged-in, etc.).

The focus of this chapter is on the metadata at level one. Hence, through the rest of this chapter, the term "metadata" is used when referring to the level-one information and not, for example, when talking about the raw log file produced by the profiling tools.

The concept of metadata for software running on embedded systems involves two parts: the metrics and the values assigned to them. The different metrics present in the set of metadata information can be classified according to their main usage. Although software metadata is defined as a whole set, different optimization tools may employ different, potentially overlapping, subsets. Figure 3.2 gives an overview of these potential overlaps. An optimization tool may take as input the values of any of the metrics, use them to transform the application and update the affected metrics with the values derived from the new behavior of the transformed application.

3.1.1 Definition and Categorization of Metadata

The first question that we have to answer while defining the concept of metadata for software applications running on embedded systems is: *Which metrics should be represented with metadata?* Any information regarding the behavior of an application that could potentially be used by *practically any* optimization tool must be included.

For software applications this mainly concerns their resource requirements (memory footprint, memory bandwidth, cycle budget, dedicated hardware needs, etc.), but also any applicable deadlines, dependencies on other software modules, events that trigger specific behavior, etc. In some cases, the metadata needed by the optimization tools can be extracted from the profiling information (the raw data) in a straightforward manner, but in most other cases more elaborate extraction techniques must be employed. Although the metadata metrics may cover all the relevant aspects of application behavior, the focus of this work is on the analysis of the memory behavior of the applications.

Figure 3.3 shows a complete view of the structure of the software metadata for the dynamic memory behavior of embedded systems as proposed in this chapter. This figure includes the semantical interconnections between different categories, such as inheritance, composition or association. Metadata information embraces aspects as distinct as the number of accesses to a variable or the relationships between execution scopes and the dynamic data types that are accessed from them. Some optimization techniques require that the metadata is defined for each dynamic data type (DDT) in the application. In particular, DDTs are shown in the figure both as abstract entities and as concrete implementations for sequences and trees, but any other dynamic data type could be added to the schema.

The different metrics contained in the metadata can be classified according to their main usage. The top level entity in the metadata is the Control Flow. This entity holds information about the dynamic data (Dynamic Data entity) that are accessed within each scope in the form of individual memory accesses and data block transfers (Block Transfer entity). These accesses happen into actual physical memory locations that are represented by the Pool entity, in which the instances of the various dynamic data are allocated. Additionally, the Control Flow entity contains aggregated information in the form of access and allocation histograms, requested bandwidth and frequency of accesses per allocated byte.

The Dynamic Data entity contains information specific to each dynamically allocated variable or dynamic data type, such as number of reads, writes or maximum required memory footprint. The type of this information is the same for variables and structured data types (however, the concrete values will be different for each case). The metadata structure reveals also the associations between the dynamic data entities and their concrete instances (Dynamic Data Instances entity). Similarly, the relation from both the data entities and their instances to the operations that are performed on them is presented in the Dynamic Data Operations entity, if applicable: for instance, the variable entity has no operations, whereas sequences or trees have several different operations such as Add, Remove, etc. The information regarding the dynamic data structures and variables of the application is presented in three levels of abstraction: aggregated at the data type level, explicit for each concrete instance of each data type or variable, and specific to the operations performed on them.

Each Variable or Dynamic Data entity is associated to a pool where its concrete instances are allocated through the system's dynamic memory manager. Therefore, the Pool entity aggregates the allocation, access, frequency per byte and bandwidth information for all the instances of the variables and dynamic data types that are

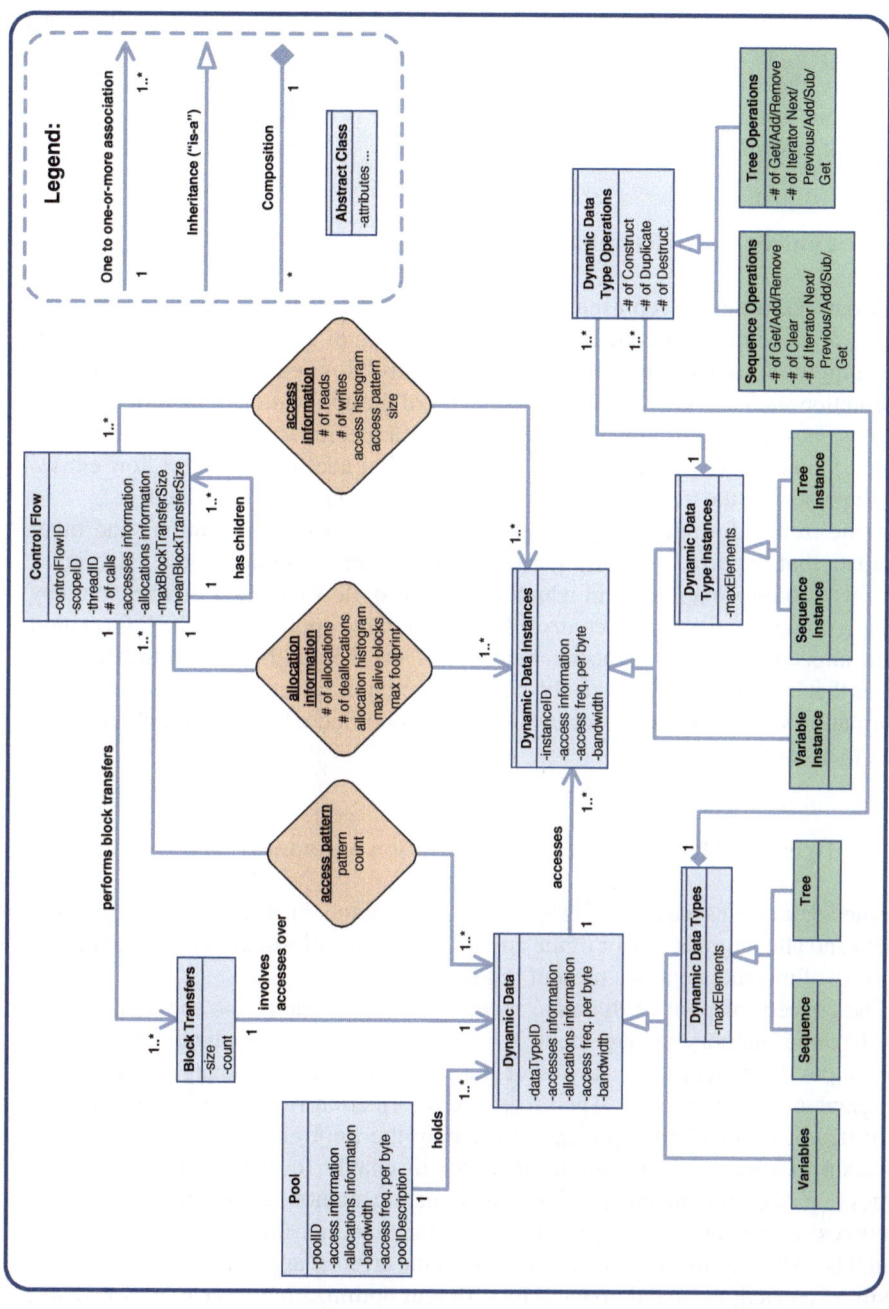

Fig. 3.3 Proposed structure for the software metadata needed to capture the characteristics of the dynamic memory behavior of applications running on embedded systems

created inside it. The Pool entity contains, in addition, metadata information regarding allocations and accesses and, more importantly, a description of the structure of the pool, namely the algorithms and data structures that are needed to perform the book-keeping of the memory.

Finally, the information about potential data block transfers (such as the size and the number of times each individual block transfer is executed) is encapsulated in the Block Transfer entity.

3.1.1.1 Control Flow Metadata

The Control Flow entity represents the control flow of the application and can be seen as a set of invocations of code locations. The purpose of this entity is to encapsulate a meaningful portion of the main application control flow, enabling the extraction of information such as the dynamic data behavior of the application at specific instants. There is one entry that is always present and gets a specific identifier to represent the main() function of the application. Any number of additional control flow entries can be present in the metadata information, each one uniquely identified.

Aside from information regarding the different accesses, allocations and other metadata that occurred within a particular section of the control flow, each entry contains also information about which other control flow entries it invokes, thereby giving a complete view of the control-flow graph of the application. The information on the allocation or memory access behavior of any thread (through all the different scopes that it activates) can be easily extracted from this information if needed. Other information such as deadlines or cycle budget is not included here to adhere to the main focus of this chapter.

3.1.1.2 Dynamic Memory Access and Allocation Metadata

The metadata entries introduced in this subsection are specific to the dynamic memory access and allocation behavior of the applications. This information is relevant to the majority of metadata entities present in Fig. 3.3.

The concept of pool of dynamic memory [171] is crucial for applications that use dynamic memory management. As the behavior of the applications becomes more input dependent, the analysis of the dynamic memory management subsystem gains more importance. Therefore, the information regarding the behavior of the different pools of the application (namely the Pool entity) is essential in the metadata of software applications in order to enable further optimization techniques (e.g., dynamic memory refinement, dynamic memory assignment on memory resources, access scheduling, etc.). It is important to notice that the pool information cannot be extracted from the analysis phase of the original application code itself, but it will be created, used and updated by different optimization tools (e.g., for DMM optimization) that will use the metadata information to communicate and pass constraints between them. The developed tools can extract, on the contrary, information such as the number of allocations and de-allocations, the maximum number of objects that are allocated at the same time and the histogram of allocations along time.

The Variable entity contains the information related to accesses to dynamic variables, with a granularity down to every dynamic variable declared in the application. Furthermore, it is possible to extract information regarding the maximum length, mean length and number of all the data block transfers that are performed in a Control Flow entity (be it a scope or thread), which is different than the number of individual memory accesses. This information, held in the Block Transfers entity for every individual potential block transfer, offers also details on which data types the transferred data *belongs to*, the size of the transfer and the number of times such block data transfers occur.

3.1.1.3 Dynamic Data Type Metadata

The concept of *DDT information* appears in Fig. 3.3. However, it is actually not a part of the metadata information of any application; rather, it is a conceptual interface to place the information related to any dynamic data type such as, a sequence (vector), a list or a tree. As an example, the figure shows the information that would be associated to the sequence and tree data types (in the Sequence and Tree entities).

Different data types have different associated information that we do not show here for the sake of brevity. Correspondingly, each concrete instance of a given data type has associated information (for instance, the Sequence and Tree Instances entities). Additionally, the Sequence and Tree Operations entities enumerate all the operations for which information can be collected on such data types. For any other data types not presented here, a new set of information for the data type, each instance and the operations would exist. However, the conceptual role that this information would take in the metadata structure is identical to the role of the related sequence or tree parts.

3.2 Metadata Mining

The starting point for the metadata extraction is the application source code. The raw information about the behavior of the data types is gathered through an extensive profiling at the dynamic data type level that triggers the application with representative input data sets. Then, the analysis phase uses these raw data in order to extract the values for the metadata metrics. Once the metadata values are defined, different tools can be linked in a pipe-line manner where they use as input the current metadata values generate an updated version of it that can be used by the next one.

Identification of relevant input data sets is crucial to obtain meaningful profiling information because the behavior of dynamic applications is heavily influenced by the nature of the input. Therefore, the value of the metrics has to be calculated for each different instance of input data. In many cases, it will be possible to abstract and make a single characterization because the values of the different metadata metrics will be the same, but in some other cases different sets of metadata values will be needed. Therefore, a mechanism to identify at run-time the characteristics of the current input instance and consequently the right set of metadata (to select the right optimizations) such as the one presented in [68] may be used in addition to the techniques proposed here.

3.2.1 Raw Data Extraction Through Profiling

The profiling step has to collect information related to the dynamic data behavior of the application so that later analysis can extract the proper metadata out of this information. Namely, the profiling step has to extract information on: a) allocation and de-allocation of dynamic memory; b) memory accesses (reads and writes); c) operations on dynamic data types; d) control-flow paths that lead to the locations where these operations are performed; and e) identifiers of the threads that perform these operations.

In general, profiling information can be obtained through two main ways: binary-code interpretation and source code instrumentation. Approaches that work directly on binary code do not require modifications to the source code, but usually the information that they extract cannot be easily related to the variables and DDTs present in it. By contrast, approaches based on instrumentation of the source code are usually able to provide information that can be related to concrete elements and locations in the source code. In this second category, it is again possible to distinguish between automated approaches (the compiler or other tool, which may even work in the same framework of the compiler, introduces the instrumentation) and manual ones (the programmer modifies the source code to introduce more specific instrumentation). For this proposal, we have opted for a semi-automated, *type-based* approach as first presented in [47]. With this solution, the designer has to annotate the types of the variables that are relevant, but the compiler takes automatically care of annotating all accesses to them through the source code. An important advantage of this choice is that the compiler ensures that no access to any of the instances of the selected data types is overlooked.

The profiling step gathers information including memory accesses, memory allocations and de-allocations, accesses to scalar variables and specific instances of the dynamic data types of the application (uniquely identified by a numeric identifier), changes of scope and the threads that perform each of the previous events. The extracted raw information keeps a sense of order, which means that aggregated constructs such as histograms of memory footprint variations at different time instances can be created.

3.2.1.1 Description and Use of the Profiling Library

In this section, we present a technique for profiling applications written in C++. However, the metadata concepts are independent of the programming language and are still valid for other (imperative) languages, provided that equivalent profiling methods are available. The type annotation is performed through the use of templates, to wrap types and give them a new type, and operator-overloading, to capture all the accesses to the wrapped variables. Templates are compile-time constructs that describe the general behavior of a class (or function) based on type parameters, thus they are sometimes known as "parameterized types". When the class template is instantiated with the desired type, the compiler generates the correct instructions to deal with that type.

The profiling library used in this chapter consists of several class templates designed to be orthogonal[1] and composable, where each of them logs different information. Additionally, the library contains a set of auxiliary classes for information formatting and recording. To keep the scope of this work under focus, we present here only the basic elements needed to understand the structure of the library.

The first auxiliary element in the library is the logger infrastructure. The following code fragment shows the basic structure of a logging class that writes a binary record for every application event:

```cpp
class DMMLogger {
    enum LogType {
        LOG_VAR_READ = 0,
        LOG_VAR_WRITE = 1,
        ...
        LOG_MALLOC_END = 5,
        ...
    };

public:
    inline static void log_read(const void * addr, const unsigned int id,
                        const size_t sz) {
        write_header(LOG_VAR_READ, 4*sizeof(unsigned int)) << id <<
            addr << sz << (unsigned long)pthread_self();
    }

    inline static void log_malloc_end(const unsigned int id,
                        const size_t sz, const void * addr) {
        write_header(LOG_MALLOC_END, 4*sizeof(unsigned int)) << id <<
            addr << sz << (unsigned long)pthread_self();
    }

private:
    DMMLogger & write_header(LogType logType, unsigned short logSize) {
        unsigned short logType_s = (unsigned short) logType;
        if (logFile_ != NULL) {
            fwrite(&logType_s, sizeof(unsigned short), 1, logFile_);
            fwrite(&logSize, sizeof(unsigned short), 1, logFile_);
        }
        return *this;
    }

    DMMLogger & operator<<(const unsigned int num) {
        if (logFile_ != NULL)
            fwrite(&num, sizeof(unsigned int), 1, logFile_);
        return *this;
    }

    DMMLogger & operator<<(const void * addr) {
        if (logFile_ != NULL)
```

---

[1] "Orthogonal" and "independent of each other" are used interchangeably in the text.

```
            fwrite(&addr, sizeof(unsigned int), 1, logFile_);
        return *this;
    }

    FILE * logFile_;      // Initialized in the constructor.
};
```

Additional methods, such as `log_write`, `log_malloc_begin`, `log_free_begin`, `log_free_end`, `log_scope_begin`, `log_scope_end`, `sequence_get`, `sequence_add`, `sequence_remove`, `sequence_clear`, `map_get`, `map_add`, `map_remove` or `map_clear`, can be easily constructed in an analogous way. The set of entries in the `logType` enumerated type has to be enlarged correspondingly.

A simple class that exports `malloc` and `free` methods with implicit logging capabilities is also needed. It can be easily implemented with code similar to the following fragment:

```
template <int ID, typename Logger = DMMLogger>
class logged_allocator {
public:
    inline static void * malloc(const size_t sz) {
        Logger::log_malloc_begin(ID, sz);
        void * ptr =::malloc(sz);
        Logger::log_malloc_end(ID, sz, ptr);
        return ptr;
    }

    inline static void free(void * ptr) {
        Logger::log_free_begin(ID, ptr);
        ::free(ptr);
        Logger::log_free_end(ID, ptr);
    }
};
```

During the actual phase of source code instrumentation, the designer uses the following class templates:

• "scope:" This is the class template used to mark different sections of the control-flow. The following code fragment illustrates its implementation:

```
template <class Logger = std_scope_logger>
class scope {
public:
    scope(std::string name)
        : name_(name)
    {
        Logger::log_scope_begin(name_);
    }

    ~scope() {
        Logger::log_scope_end(name_);
    }
private:
    std::string name_;
};
```

The designer can declare an instance of this class template at the beginning of any interesting section of code and the compiler-generated constructor and destructor will generate the profiling tokens. Then, the analysis tools may determine which portions of code were executed, the sequence of activations and, most importantly, which portions of the code were responsible for the different profiling events. Moreover, the scope template can be combined with thread identification to obtain the complete control-flow graph of the application for an input set and to relate the rest of the events (allocations, accesses to dynamic data types) to specific portions of code executed by specific threads.

- "allocated:" Class instances are normally created and destroyed through the new and delete operators. This class template automates their overloading to generate profiling tokens. Typically, profiling tokens are generated for each of the operations and the requests are forwarded to the system allocator through the malloc and free functions:

```
template <class Allocator>
class allocated {
public:
    void * operator new(const size_t sz) {
        return Allocator::malloc(sz);
    }
    void operator delete(void * p) {
        return Allocator::free(p);
    }
    void * operator new[](const size_t sz) {
        return Allocator::malloc(sz);
    }
    void operator delete[](void * p) {
        return Allocator::free(p);
    }
};
```

In order to log all the dynamic memory events related to the instances of a class (or structure), the designer would declare it as follows:

```
class NewClass : public allocated<logged_allocator<1> > {
    ...
};
```

No other lines in the source code of the class need to be modified.

- "var:" This class template wraps the declaration of individual variables or class attributes and generates a profiling token for each memory access to them. Additionally, it derives from the "allocated" template, thereby also giving allocation and de-allocation logging of the wrapped variable. The following is an excerpt of the template implementation:

```
template <typename T,  // Type of the wrapped object
          int ID,       // Id used in the profiling tokens
          class Logger = DMMLogger,
          class Allocator = logged_allocator<ID, Logger> >
class var : public allocated<typename Allocator> {
public:
    T data_;

    // Constructor
    template <typename T2>
    var(const T2 & data)
      : data_(data)
    {
        Logger::log_write(&(data_), ID, sizeof(T));
    }
```

```
    // Assignment operator from basic type
    template <typename T2>
    var & operator= (const T2 & data) {
        data_ = data;
        Logger::log_write(&(data_), ID, sizeof(T));
        return *this;
    }

    // Assignment operator from wrapped type
    template <typename T2, int ID2>
    var & operator= (const var<T2, ID2> & other) {
        Logger::log_read(&(other.data_), ID2, sizeof(T2));
        data_ = other.data_;
        Logger::log_write(&(data_), ID, sizeof(T));
        return *this;
    }
};
```

This template can be used to wrap variables of any basic type, including pointers and class attributes. To profile accesses to the attributes of a class or structure, it is necessary to go inside that class and wrap all the relevant attributes. If any of the attributes are also structures, then the technique can be applied recursively until the basic types are reached:

```
var<int, 1> var1;
var<int, 2> * pointer1 = new var<int, 2>;

struct A {
    var<int, 3> foo;
};
struct B {
    A a;
    var<int *, 4> bar;
};
```

A special consideration is needed when wrapping pointers. In the following code fragment, the first declaration means that only accesses to the pointer quux, but not the accesses to the integers that are pointed to, are logged. However, when both accesses are of interest, the pattern given for the definition of baz, which is a combination of the previous patterns, should be used instead:

```
var(int, 1) * quux;
var(var(int, 2) *, 3) baz;
```

Using the "var" class template, the designer does not need to worry about identifying all the positions in the source code where the variables are accessed as the compiler will ensure that they are properly logged.

- "vector:" This class template replaces the vector from the STL [145] in order to profile sequence usage behavior at the dynamic data type abstraction level. This information can be used, for instance, to decide what is the ideal data type implementation, as shown in [16]. The implementation of the template matches that of the STL one from a logical point of view, but adds logging at the right places to give a high-level view of the usage pattern of the data structure (e.g., mostly insertions, linear traversals, random accesses, etc.). The declaration of a vector with this template is straightforward:

```
vector<int, 1> aVector;
```

Finally, the designer can combine the "vector" and "var" templates to profile also the accesses to the actual data items inside the sequence.

The "allocated," "var" and "vector" templates are parameterized with a unique identifier that is recorded in the profiling tokens; hence, the profiling information can easily be linked to the original source-code. The "scope" template is parameterized with a human-readable identifier for easier tracking.

3.2.2 Analysis Techniques for Metadata Inference

Once the profiling information has been extracted from the application, several analysis steps can be applied to extract and compute the relevant metadata metrics that will be used by the various optimization tools to reduce energy consumption, memory accesses and memory footprint. Each of these optimization tools will use specific parts of the metadata set. Therefore, the metadata information can be seen as structured in different, potentially overlapping, slices of interest. In this section, we present several techniques that can be used to extract different portions of the metadata information.

The analysis process is structured as a set of objects that perform specific analysis tasks. The main driver reads every packet from the log file produced during the profiling phase and invokes in turn all the analysis objects to process it. After all the analysis objects have had the chance to compute the information related to the current packet, the main driver moves forward to the next one. Due to the way that our profiling information is gathered, it is not meaningful to have absolute time-stamps, as the time that it requires to profile dominates (thus, clobbering) the run-time of the application.[2] Therefore, our timing measure is defined in terms of the amount of profiling packets (of a specific type, such as allocation or access packets) that have passed since the beginning of the execution. However, this is a good measure for the type of analysis proposed here, due to the fact that all the metadata metrics, as well as the analysis and optimization methods, deal with dynamic memory accesses and not with the computation time. As a result, the timing is based on events that alter the state of the dynamic memory (e.g., allocations) or that define milestones for the memory subsystem (e.g., number of accesses).

---

[2] This is an important factor when using the profiling techniques proposed in this chapter. The fact that every access to every dynamic object in the application is logged introduces a high overhead on execution time. The case study shown in this chapter is based on network traces that were collected off-line and thus the whole activity of the system can be replayed using a notion of "virtual execution time" that is independent from the actual "wall time" that it takes to run the experiments. However, for cases where the behavior of the system cannot be analyzed without a real notion of time (i.e., systems where the behavior changes subject to real-time conditions), more advanced profiling techniques out of the scope of this chapter would be needed. For example, the memory subsystem could be augmented during development with a parallel structure to log the different data accesses in real-time, as presented in [40].

Algorithm 3.1 shows the work performed by the analyzer of accesses to variables. Whenever a read or write event is found in the log file, the global read or write counters (for the DDTs) of the application and the current scope and thread are updated. Using the address of the memory access, it is possible to find the concrete instance of a dynamic data type that was being accessed. This information allows updating also the number of accesses of the corresponding dynamic data type and DDT-instance.

Algorithm 3.1 Access analysis

```
// event.type is variable read or write
process(event) {
  Update application reads/writes counter
  Update reads/writes counters of current control-flow point
  FindBlockOfAddress(event.address) =>
    Update block reads/writes counter
    Update DDT & DDT-instance reads/writes counter
}
```

Another important analysis target is the allocation behavior as such information can be used to design highly-tuned application-specific dynamic memory managers. To enable this optimization, we identify the number of allocations per block size, the different block sizes and the number of accesses per block size. With this information it is possible to automate the exploration of application-specific dynamic memory managers, as demonstrated in [105]. Additionally, the accesses to variables encountered between `MallocBegin` and `MallocEnd`, or between `FreeBegin` and `FreeEnd`, can be used to evaluate the overhead of different dynamic memory managers. Finally, these data can also be used to detect memory leaks. Algorithm 3.2 shows how all this information is extracted.

Identifying block transfers amongst the individual, profiled, memory access events, enables the application of data transfer optimizations. For example, dedicated hardware resources (i.e., DMA engines) can be used to perform the data transfers that are long, saving computing cycles for the main processing elements. Several trade-offs exist when deciding the kind of optimizations that should be applied; for instance, in [135] a technique to selectively perform data transfers using a DMA module in embedded systems is presented. We propose to define a unique data transfer as a set of strictly consecutive data accesses to the same instance of the same data type, performed by a given thread. With this definition, we use Algorithm 3.3 to identify the data transfers of the application.

An additional optimization that is enabled by the identification of patterns in the accesses to data elements is the placement of those data elements that are usually accessed consecutively into *burst-friendly* memories (e.g., DRAMs), thereby reducing the number of page activations and mode changes in the memory elements. We propose Algorithm 3.4 to identify access patterns in the accesses to dynamic data types. This algorithm is responsible of building a list of access patterns that occur during the execution of the application. Furthermore, each access pattern is decom-

Algorithm 3.2 Block information gathering

```
process(event) {
  case event of
    AllocEnd ->
      Create new live-block with event.size, event.address
        and the control-flow information from
        activeControlFlows.get(threadId).top()
      Increase allocation count for event.size
    DeallocEnd   ->
      FindBlockOfAddress(event.address) =>
        Increase de-allocation count for event.size
        Add number of accesses to accesses for blocks
         of size event.size
        Destroy live-block
    VarRead or
    VarWrite     ->
      FindBlockOfAddress(event.address) =>
        Increment read or write counter for this live-block
}
finalize() {
  forall block ∈ live-blocks {
    Report memory leak for this block and where it was
      allocated
  }
}
```

posed into smaller, more primitive ones. In this way the designer has a complete view
in a multi-granular manner of the access patterns that characterize the application.
This algorithm can be made more efficient by using a maximum *lookback* window.
Through experimentation we have found that, for dynamic data behavior, a window
of 100 elements identifies the same pattern as one that is not window-limited, while
analysis times remain bounded.

Algorithm 3.3 Data transfers identification

```
process(event) {
  case event of
    DeallocEnd ->
      FindBlockOfAddress(event.address) =>
        Record last active transfer for the block
        Destroy block
    AllocEnd ->
      Create new block with event.address
    VarRead or
    VarWrite ->
      FindBlockOfAddress(event.address) =>
        if consecutive access for event.threadID and
            same direction (read/write) {
          Update active transfer with event.address
        } else {
          Record last active transfer (if any)
          Create new transfer with event.address
        }
}
```

While the previously described algorithm is very exact and can give a very specific and concrete view of the behavior patterns, in the case of very dynamic behaviors, Exact Matching patterns will not be detected. In this case, we propose Algorithm 3.5, which extracts non-exact patterns that do not keep track of the number of repetitions inside them, and thus of the number of times that events are repeated. In this way each pattern does not hold fine grain information regarding the patterns it is composed of. A motivational example for this is given here:

Example: Pattern Matching

Given the following stream of events: AAAABAAABAAABAAAB, where A means accesses to variable$_1$ and B means accesses to variable$_2$, the following patterns would be deduced:
Exact Matching: 4*[3*[A], B]
Non-Exact Matching: [[A], B]

It seems clear that variable$_1$ and variable$_2$ are being accessed in a loop. If, however, the number of accesses to variable$_1$ varies in this loop due to dynamic, data-dependent, conditions, it is possible to have a stream such as: AAAABAAAAABAAABAB. Then, the pattern extracted would be:
Exact Matching: [4*[A], B, 5*[A], B, 3*[A], B, A, B]
Non-Exact Matching: [[A], B]

Non-exact pattern-matching can be beneficial for applications demonstrating dynamic behavior, even though it offers less information than exact pattern-matching.

These two algorithms are not limited to accesses; they can be used for pattern matching of any of the profiling events, for instance allocation or scope-events. In conclusion, the two algorithms serve different purposes: The exact algorithm gives the dominant behavior of the application, but may miss out on less stable patterns; the non-exact algorithm gives a global look of what sort of patterns to expect.

Gathering information on the statistical behavior of different sequences in the application enables to discern the average number of elements in the sequences of each specific type. The algorithm to extract this information is straightforward; thus, it is not presented here. This information and the histogram of the operations invoked for each sequence is crucial not only for identifying the dominant data types, but also to decide the sort of dynamic data types that should be used [16].

3.3 Case Study: Integrated Example on DRR Scheduling

Here, we present an integrated approach to extract and exploit the characteristics of applications running on embedded systems by means of defining a common *software metadata*. Therefore, it is important to present also an integrated example where different optimization methods are applied on a single application. Moreover, this

Algorithm 3.4 Exact meta pattern matching algorithm

```
Type Pattern a = case Kind of
  Element => a
  Pattern => record {
    count :: Integer,
    pattern :: List (Pattern a)
}

current :: List (Pattern a)
initialize() {
  current = new List;
}
process(event) {
  case event of
    type -> patternize(new List(Element type));
}
patternize(segment) {
  n = segment.length;
  if (n > windowsize || current.length == 0) {
    current.append(segment);
  }
  if (current.get(-1).pattern == segment) {
    // Segment fits a pattern
    // Increase pattern count
    current.get(-1).count = current.get(-1).count + 1;
    // Create a new segment to look for
    segment = new List(current.get(-1));
    current.dropLast(1);
    // Start fresh with higher-order pattern
    patternize(segment);
  } elseif (current.range(-n, -1) == segment) {
    // Segment fits another segment
    // Drop the segment in current
    current.dropLast(n);
    // Create a new segment to look for
    segment = new List(Pattern 2 segment);
    // Start fresh with higher-order pattern
    patternize(segment);
  } else {
    // Try to look for a bigger match
    segment.prepend(current.get(-1));
    current.dropLast(1);
    patternize(segment);
  }
}
```

example shows how software metadata enables relevant optimizations resulting in significant gains in energy consumption and/or memory footprint. A network TCP/IP-like stack including Deficit-Round-Robin (DRR) scheduling of outgoing packets is used as the case study application to illustrate the synergistic effect obtained when the software metadata is used by several optimization tools on the same application.

Algorithm 3.5 Non-exact meta pattern matching algorithm

```
Type Pattern a = case Kind of
  Element => a
  Pattern => List (Pattern a)

current :: List (Pattern a)
initialize() {
  current = new List;
}
process(event) {
  case event of
    type -> patternize(new List(Element type));
}
patternize(segment) {
  n = segment.length;
  if (n > windowsize || current.length == 0) {
    current.append(segment);
  }
  if (current.get(-1).pattern == segment) {
    // Segment fits a pattern
    // Create a new segment to look for
    segment = new List(current.get(-1));
    current.dropLast(1);
    // Start fresh with higher-order pattern
    patternize(segment);
  } elseif (current.range(-n, -1) == segment) {
    // Drop the segment in current
    current.dropLast(n);
    // Create a new segment to look for
    if (n == 1) {
      // Keep current segment, namely of one item.
    } else {
      // Generate a pattern out of multiple items.
      segment = new List(Pattern segment);
    }
    // Start fresh with higher-order pattern
    patternize(segment);
  } else{
    // Try to look for a bigger match
    segment.prepend(current.get(-1));
    current.dropLast(1);
    patternize(segment);
  }
}
```

3.3.1 Goal and Procedure

The final goal of this example is to optimize the implementation of the system at three different key points: dynamic data type, dynamic memory management and block transfers of dynamic data. The steps performed are:

1. Extraction of the software metadata from the original application(s) using the profiling and analysis techniques presented in this chapter.
2. Optimization of the dynamic data types to reduce the number of memory accesses and the memory footprint of the dynamic data structures used in the application (e.g., linked lists, double-linked lists).

3. Optimization of the dynamic memory management to support a more efficient implementation of the `malloc()` and `free()` operations. After this step, the number of memory accesses performed to manage the dynamic memory is minimized. Correspondingly, the total memory footprint required by the dynamic memory manager to serve all the demands of the applications is also minimized (i.e., by reducing the internal and external fragmentation caused by the manager itself [171]).
4. Optimization of the block transfer behavior of the application to exploit the DMA resources for transfers of blocks of dynamic data (as opposed to the transfer of static arrays or variables). The result of this step takes into consideration the effect of the concurrent accesses from the processor and the DMA on the banking scheme of DRAM memories.

The memory behavior of the application is evaluated after each step, in order to compare the impact of each decision. In the end, the application is simulated to attest the accumulated effect of all the optimizations.

3.3.2 Description of the Case Study Application

A TCP/IP-like network subsystem is used throughout this example as the case study application. This system is organized in several threads that communicate through asynchronous FIFO queues: the output of one thread is the input for another one. A more detailed explanation of this application can be found in [135]. Basically, the application is divided into the following modules:

- Packet injection. A collection of real (wireless) network traces [73] is used to generate the packets that are fed into the system.
- Packet formation. The TCP/IP header is added to the data in order to build a complete packet.
- Encryption (DES). This module is bypassed by packets of sessions that do not require encryption.
- TCP Checksum.
- Scheduling and Quality-of-Service management. Deficit-Round-Robin (DRR) is a network fair scheduling algorithm commonly used for scheduling according to available bandwidth [148]. This algorithm has been implemented in various switches (e.g., Cisco 12,000 series).

These subsystems form the basis of a simplified network stack. Due to the fact that we use network traces collected from the wireless access points of a university campus [73], and not from the actual devices, reproducing details like packet retransmission or rate control becomes almost impossible; hence, we stick to the simplified description presented before. As the analyzed application is multi-threaded, many packets are alive at the same time during execution; thus, memory accesses are performed concurrently from multiple threads. Multi-threading eventually means that

Table 3.1 The SDRAM is modeled according to Micron PC100 specification assuming CL = 2 and a system and processor clock of 100 MHz

Energy/access	3.5 nJ
Energy activate/precharge	10 nJ
CAS latency	2 cycles
Precharge latency	2 cycles
Active to read or write	2 cycles
Write recovery	2 cycles
Last data-in to new read/write	1 cycle
Max burst length	1,024 words

memory accesses from different threads interleave in a fine-grained way and the overall behavior cannot be described from the independent behavior of each thread. Therefore, the whole system must be optimized instead of each thread independently.

The target architecture for this system consists of a processing element connected to an SRAM module and to an external DRAM module (see Table 3.1 for a description of their working parameters), as presented in [135]. A Direct Memory Access (DMA) engine takes care of data movements between the external memory and the internal, more efficient one, and from any of the memories towards the external devices (i.e., hardware buffers in the network adapters). The processor can access both memories directly; however, access to the external DRAM has to be coordinated with the DMA to avoid incurring unnecessary energy and latency penalties due to page-level interferences.

3.3.3 Profiling and Analysis

The case study application is instrumented and profiled, and the extracted information analyzed as explained in Sect. 3.2. After this step, performed once for all the optimizations, the initial metadata is ready. The extracted information includes:

- For each dynamic data structure, the number of operations of each type executed, the number of memory accesses needed to accomplish them and the total memory footprint.
- For the dynamic memory manager, the number of memory accesses executed to manage all the free and used blocks of dynamic memory, and the total amount of memory actually used to serve all the petitions from the application. The total usage of memory depends on the internal and external fragmentations, which are caused by the manager when subject to the application behavior.
- Finally, the number of accesses to each dynamic object and the access pattern of each thread in the system.

This information allows identifying the most relevant dynamic data types in the application. The first one corresponds to the packets that are created to be sent. Their huge number demands high efficiency from the dynamic memory manager. The next most relevant data structures are the list of nodes and the queue of packets built by the DRR module. The list of nodes represents the hosts to which a connection is active, i.e., there is an entry for each destination host for which there are packets waiting to be sent. Each entry in the list contains the queue of packets waiting to be sent towards this destination. Both data structures are dynamic because their number of elements varies at run-time; thus, they are good candidates for the data type optimization techniques. Finally, the last of the relevant dynamic data structures used in the application is the asynchronous FIFO queue that the threads use to pass messages between them. This data structure makes heavy use of synchronization primitives, which are out of the scope of this work. Nevertheless, the best implementation of this data structure may be determined in a straightforward way because it represents a very regular access pattern.

Each of the next three optimization steps relies in the metadata extracted at this point and updated by the corresponding tools.

3.3.4 Dynamic Data Type Refinement—DDTR

The first optimization applied in this example is the refinement of the dynamic data structures of the application (DDTR). After the profiling and analysis steps, the metadata entities Dynamic Data Types, Dynamic Data Types Instances and Dynamic Data Types Operations contain information on the type and number of operations performed for each data type. Most importantly, this information is also available for their concrete instances, which makes it possible to study and optimize each of them according to the operations performed more frequently. The profiling and analysis methods allow to differentiate between the accesses issued to the data structures themselves, and the accesses to the internal structures of the implementing data types. This differentiation makes it possible to identify the most relevant data types for optimization independently of their initial implementation. Furthermore, our optimization techniques focus on optimizing the overhead imposed by each data structure, but they cannot reduce the number of accesses that the application algorithms execute. In order to reduce the number of accesses issued to the dynamic data structures (in contrast to the number of memory accesses actually performed on the memory subsystem), optimizations at a higher abstraction level that are out of the scope of this chapter would be needed.

The extracted information reveals that the two dynamic data instances that represent most of the application accesses are the list of nodes in the DRR algorithm (dynamic structure "A") and the queue of pending packets for each of the nodes (dynamic structure "B"). Using the methods explained in Chap. 4, we choose the best option for their respective implementations. Although running an exhaustive exploration of all cases is not needed to get the optimal solution, in this experiment we ran a complete sweep of all the combinations to perform an additional comparative test and validate the optimization approach. For each of the dynamic data structures, one of the following implementations may be chosen (more details on each of the implementations can be found at [108]):

1. Array of pointers
2. Array of objects
3. Single-linked list
4. Double-linked list
5. Single-linked list with roving pointer
6. Double-linked list with roving pointer
7. Single-linked list of arrays
8. Double-linked list of arrays
9. Single-linked list of arrays with roving pointer
10. Double-linked list of arrays with roving pointer.

Trees are not represented in this hierarchy, but they can be built using a combination of these data structures. For more efficient exploration of trees, a further study of the possible interfaces that a tree may expose would be required and this is out of the scope of this book. Since most tree implementations are rather ad-hoc, and from a high point of view different tree types have different properties, it would be necessary to first classify the different tree data structures before trying to define commonalities in their interfaces. Some tree data structures, such as ordered trees, fundamentally behave as sets. Other tree data structures have more fundamental encoding in the tree data structure, such as parent-child properties.

In the rest of this Section, Ai–Bj represents the combination of the implementation i for A and the implementation j for B. For example, A1–B3 represents that an array of pointers is used as implementation for the list of nodes in the DRR algorithm and a single-linked list is used as implementation for the queue of pending packets for each node.

There are ten possible implementations for each of the dynamic structures. Due to the fact that we wanted to do a full test of all cases, that we used thirteen input traces as the input set and that we executed ten repetitions to tackle with statistical variations, the total number of combinations run ascended to 13,000. However, this high number of executions was performed just to validate the optimization with an exhaustive test, but in practical cases there is no need to perform them. For this extensive experiment, the metadata for each of the dynamic data instances was collected from the metadata entities Dynamic Data Types, Dynamic Data Types Instances and Dynamic Data Type Operations, with additional information in Access Information (Fig. 3.3).

Two conclusions are drawn from the results of the experiments. The first one is the confirmation of the most efficient data structure regarding the total number of memory accesses. The second one is the analysis of the data structure that reduces most the total amount of memory (memory footprint) needed to execute the application.

Table 3.2 Comparison of the solution A3–B3 to a hypothetical "perfect" solution for each input case, in terms of memory accesses

Input set	Memory accesses A3–B3	Memory accesses optimal	Difference (%)
01	4 328 501 556	4 247 377 390	1.91
02	3 589 773 903	3 530 754 848	1.67
03	8 004 231	8 004 231	0
04	1 867 812	1 867 812	0
05	9 622 944	9 622 944	0
06	25 225 070	25 225 070	0
07	192 869 416	192 869 416	0
08	183 636 289	183 636 289	0
09	12 699 747	12 699 747	0
10	45 429 504	45 429 504	0
11	239 756 665	239 756 665	0
12	19 082 840	19 082 840	0
13	250 188 588	250 188 588	0
Avg. diff.			0.28

The selected solution is optimal in 11 out of the 13 input cases considered and the average difference is only 0.28 %

3.3.4.1   Reducing the Number of Memory Accesses

The experimental results show that the most efficient combination of dynamic data structures is A3–B3, which means that a single-linked list implementation should be used for both the list of active nodes in DRR and the queue of packets waiting to be sent for each of the nodes. This solution is the optimal for 11 of the 13 inputs that were considered. Table 3.2 shows the number of accesses required by the optimal solution for each input, and the number of memory accesses required by the selected solution A3–B3. The forth column shows the percentage difference between both. It is relevant to note that the average difference from A3–B3 to a hypothetical "perfect" solution[3] is as little as 0.28 %.

3.3.4.2   Reducing the Memory Footprint

In the case of memory footprint, a Pareto optimal solution considering also the effect on the number of memory accesses was evaluated. Table 3.3 shows the results obtained from the experiments. In this case, A3–B3 is optimal in only one of the 13 input cases. The overall difference from this solution to the optimal memory footprint solution (with very bad memory access trade-offs) is only 3.33 %. This means that, even if there is not a "perfect" solution to reduce the memory footprint

---

[3] This "perfect" solution is calculated by selecting the optimal DDT combination for each input case in isolation. Then, the configuration selected is compared against each of them to calculate how far it lies from the optimal for each input case.

Table 3.3 Comparison of the solution A3–3 to a hypothetical "perfect" solution that minimizes the memory footprint for each input case

Input set	Memory footprint A3–B3	Memory footprint optimal	Difference (%)
01	9 028 421	8 992 232	0.40
02	7 882 678	7 872 560	0.13
03	1 978 361	1 809 807	9.31
04	447 778	447 778	0
05	992 569	953 663	4.08
06	2 178 989	2 025 122	7.60
07	186 441	185 851	0.32
08	174 315	171 398	1.70
09	1 589 535	1 532 995	3.69
10	3 316 557	3 271 309	1.38
11	760 281	716 898	6.05
12	772 158	732 497	5.41
13	471 254	456 433	3.25
Avg. diff.			3.33

The average difference is 3.33 %

of the application, selecting the best compromise solution to reduce the number of accesses does not inflict a significant penalty on the total amount of memory required.

3.3.5 Dynamic Memory Management Refinement—DMMR

Once the dynamic data structures of the application have been optimized and the corresponding metadata information is updated (Dynamic Data related entities, Fig. 3.3), the optimization of the dynamic memory manager is performed, as further detailed in Chap. 6. We refer to the dynamic memory manager as an integrated set of application specific dynamic memory managers (i.e., application level DM managers compiled with each software application). If more than one application is present, then each of them gets compiled with its own customized DM manager. In any case, all the customized DM managers share common OS level services for providing big memory blocks (e.g., sbrk() and mmap()). The metadata information for this step includes the number of blocks of varying sizes that are allocated by the application for the network packets (data bodies and network headers independently), the list of destination nodes for the DRR algorithm and the queues of pending packets for each node.

Performing this step after the dynamic data structures have been optimized is important because the number of memory allocations and de-allocations will not change anymore due to implementation modifications. Every time a new instance of any of these dynamic data types is created, the dynamic memory manager must

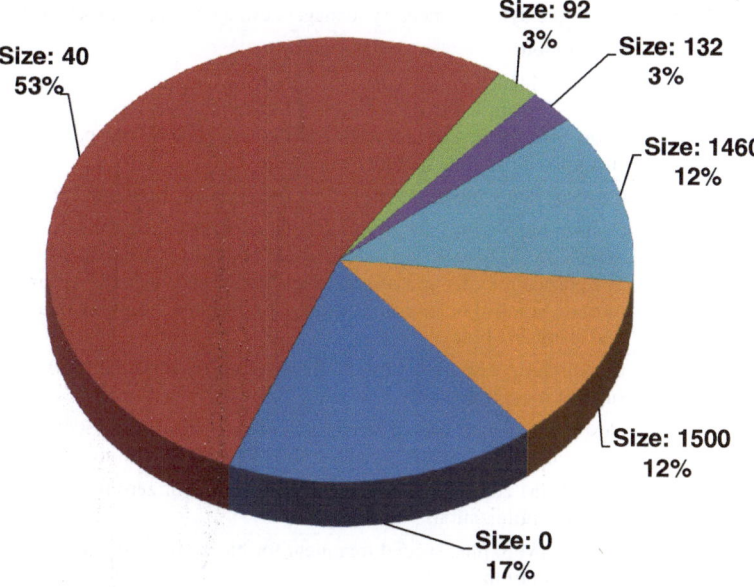

Fig. 3.4 Frequency distribution of "popular" allocation sizes for all the input cases

search for a suitable free block, i.e., a block of sufficient size that spans over a range of memory addresses that are not currently occupied by any other object. Conversely, when an instance is destroyed, the memory space that it was using must be reintegrated to the pool of available addresses for new instances. All of these operations produce an overhead in the total amount of work that the system has to do. Additionally, the memory manager requires some extra memory for its own operation and the book-keeping of the whole process increases the total number of memory accesses. Therefore, the suitability of a given dynamic memory manager depends on the overhead that it imposes on the system and the extra memory needed due to the existence of internal and external fragmentation [171].

The size of the allocated memory blocks and their frequency of appearance, together with the pattern of allocations and de-allocations, determine the character-istics of the most suitable memory allocator. These numbers are defined in the soft-ware metadata of the application (Dynamic Data and Pool related entities, Fig. 3.3). Figure 3.4 shows the frequency of appearance of each allocation size in the driver application, for the most popular sizes. The figure makes it clear that the memory manager for this application must be optimized to allocate large quantities of packets from a small set of block sizes.

In order to reduce the number of memory accesses, the memory manager has to locate the most appropriate free block with the least amount of memory accesses. For this reason, a memory manager gets first free memory space from a global pool and then frees the blocks into lists of specific sizes. Additionally, in order to reduce the

Table 3.4 Configuration of the dynamic memory managers evaluated in this experiment

DMM	Description
DMM 1	Kingsley-like [125] memory manager with bins for blocks of 128 different sizes, from 8 to 16384 bytes. This popular memory manager is used throughout the rest of this section as the reference for comparison with the customized dynamic memory managers
DMM 2	Custom heap with lists of free blocks of sizes 40, 1460 and 1500 bytes
DMM 3	Custom heap with lists of free blocks of sizes exactly 40, exactly 1460, exactly 1500, up to 92, up to 132, up to 256, up to 512 and up to 1024 bytes. The particular order and constrains of "up to" and "exactly" ensure that the most common allocation sizes require the minimum number of accesses to find a suitable block
DMM 4	Custom heap with lists of free blocks of sizes 40, 1460 and 1500 bytes, plus support for splitting and coalescing
DMM 5	Custom heap with lists of free blocks of sizes 40, 1460, 1500, 92 and 132 bytes (in order of search priority), plus support for splitting and coalescing
DMM 6	Like DMM 2, plus special treatment for blocks of zero bytes (application specific optimization)
DMM 7	Like DMM 3, plus special treatment for blocks of zero bytes (application specific optimization)
DMM 8	Like DMM 4, plus special treatment for blocks of zero bytes (application specific optimization)
DMM 9	Like DMM 5, plus special treatment for blocks of zero bytes (application specific optimization)

amount of internal fragmentation, it is also possible to create lists of free blocks for a small number of additional sizes that limit the amount of wasted memory for uneven sizes. The structure of the memory allocator is held into the `poolDescription` entry of the `Pool` entity (Fig. 3.3).

An important consideration is that the subject application does actually request blocks of zero bytes in size. This is not an error, but a consequence of the need to send acknowledgement-only packets in the absence of outgoing data when the system is receiving big amounts of data from the network. As these 0 Byte TCP segments are moving to the IP level, the segment body of zero bytes is kept and a 40-bytes TCP + IP header is added.[4]

With the previous considerations, and the methods presented in Chap. 6, the design space of the dynamic memory manager may be narrowed to a few options. For the case study application of this example, we evaluated the dynamic memory managers

---

[4] Even if the 0 Byte TCP segment is empty, according to the rules of the C/C++ programming language, an allocation of zero bytes is valid and must return a valid (not NULL) object. However, the actual size of the object is zero bytes and the application cannot access any bytes at this address. Indeed, the previous versions of the C++ standard did not require that a distinct block was returned for zero-size blocks. Based on this fact, an additional optimization may be introduced in the memory manager to ease the allocation of zero-byte blocks: the memory manager may use a fixed block to host all of these requests. Actually, this method could be used to detect application errors during the developing phase by making this special block access-protected.

described in Table 3.4. After each experiment, the information generated in the `Pool`, `Allocation Information`, `Dynamic Data`, `Dynamic Data Instances` and `Control Flow` metadata entities (Fig. 3.3) was collected to evaluate the quality of each solution.

Proposed blocks: "Special" memory block sizes should be predefined which are equal to each packet size that represents at least 10 % of the overall packet sizes. The rest of the predefined memory blocks should be power-of-two sizes up to the MTU size. In the case of the IEEE 802.11b, this means that one special predefined memory block of 40 bytes and one special predefined memory block of 1,500 bytes are present. The rest of the predefined memory blocks should have 256, 512 and 1,024 bytes, respectively. For some of the cases considered in the experiments of this chapter, we added two smaller sizes (92 and 132 bytes). Therefore, the most popular memory requests can be satisfied without any internal fragmentation and the remaining less popular requests can be satisfied with reasonable internal fragmentation. Most importantly, the design decision of having blocks of fixed sizes, predefined at compile-time, gives the performance advantage of not having to calculate the block size for each request at run-time.

Proposed coalescing and splitting of blocks: No splitting or coalescing of memory blocks should be used. As the maximum requested block size is already known (the MTU of the packets), there is no need to coalesce blocks to deal with external fragmentation. Additionally, defining block sizes that prevent most of the internal fragmentation (i.e., the internal fragmentation produced by the popular requests) reduces the need to split blocks. Most importantly, both the splitting and the coalescing mechanisms are computationally intensive, thus they may slow down substantially the allocation and de-allocation processes and may also incur a significant number of memory accesses to transform the old block sizes into the new ones.

Proposed pools: A number of pools equal to the number of the predefined block sizes should be created. In the present case, two special pools, which hold the 40 bytes and the 1 500 bytes blocks, and one pool for each additional block size are created. This lean pool organization is preferred instead of more complex ones because it allows a faster access to the specific memory pool that will service each request: in the worst case, the number of memory accesses that will be needed to find the pool that holds the block of the appropriate size is in the order of the number of pools. Finally, once the decision of not supporting coalescing nor splitting of blocks is made, the DMM does not need to support (performance-costly) movements of blocks between pools.

Proposed fit algorithms: In order to choose a pool, the Exact Fit and First Fit algorithms are proposed. Exact Fit is used only to discriminate between the special pools, while First Fit is used between the rest of them. Once a pool is chosen, only First Fit is used to choose the appropriate block. This configuration requires the least accesses in order to find the memory block to service a memory request. If the wrong combination of pool sizes and fit algorithms is made, the number of memory accesses executed by the DM manager can increase significantly leading to longer execution times and higher energy consumptions.

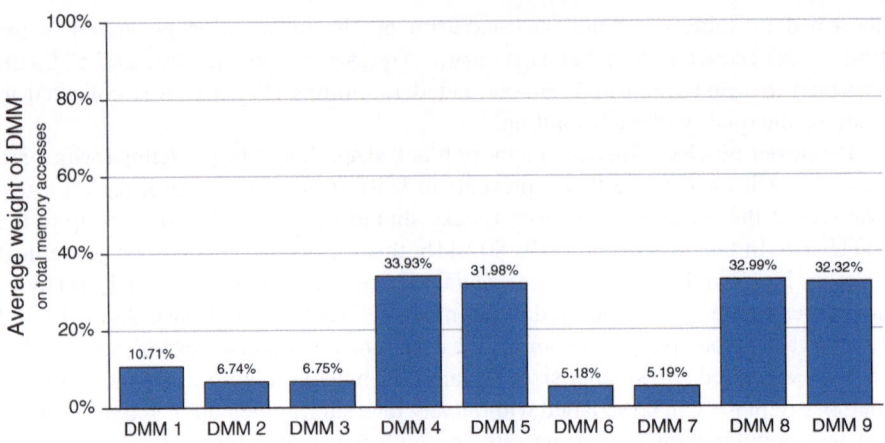

Fig. 3.5 Percentage of memory manager accesses over total application accesses

As a final consideration to motivate the relevance of the dynamic memory manager for the global performance of the complete application used in this example, Fig. 3.5 shows the weight of the memory accesses[5] due to the dynamic memory management over the total number of memory accesses, for each of the memory managers used in our experiments. The dynamic memory managers 2, 3, 6 and 7 require a relatively low number of memory accesses to perform their work. On the contrary, DMMs 4, 5, 8 and 9 introduce more than one third of the memory accesses required by the application just to manage the dynamic memory. DMM 1, the reference design, introduces a moderate 10 % of additional accesses.

Figure 3.6 shows the total amount of memory accesses needed to process each input case with the DMMs 1, 6 and 7. Then, Fig. 3.7 reveals that for the data intensive cases, where mostly big packets are sent and the work required to process the data supersedes the work needed to allocate the blocks, the number of accesses due to dynamic memory management represents less than 1 % of the total (inputs 1 and 2 in both figures). However, when the input of the application forces to send many small packets (that is, small TCP segments are sent due to the necessity of bounding the transmission delay time even under low application traffic), the number of accesses due to memory management can scale up to a 24 % of the total for the input number 8 using DMM 1. Interestingly, the overhead of DMMs 6 and 7 is approximately half of that, underlining the importance of optimizing the performance of the dynamic memory manager.

---

[5] This weight is calculated taking into account the number of memory accesses that are logged between the *MallocBegin* and *MallocEnd* profiling tokens, and between the *FreeBegin* and *FreeEnd* ones. These accesses correspond to the work that the memory manager has to do in order to handle each application request. The fraction of accesses that correspond to the overhead of doing memory management, i.e., their weight, is obtained dividing this number between the total number of memory accesses.

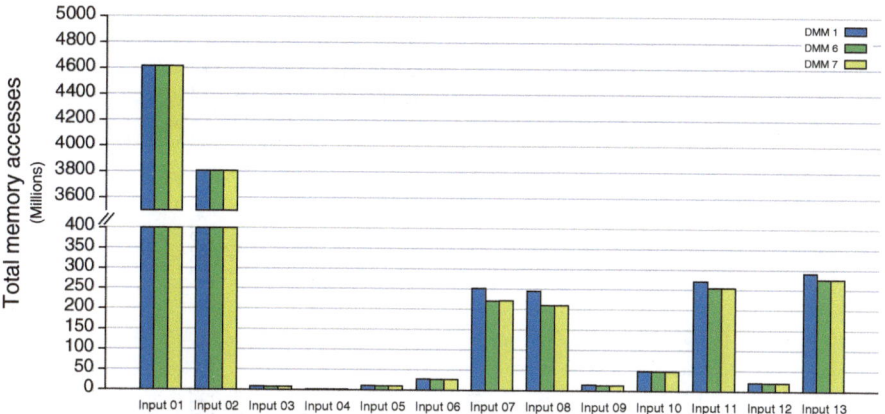

Fig. 3.6 Impact of dynamic memory management on the total number of memory accesses: total number of accesses for each input

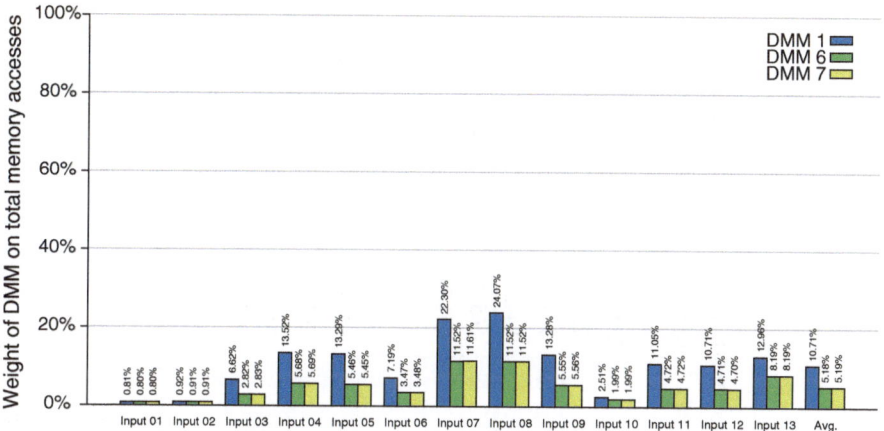

Fig. 3.7 Impact of dynamic memory management on the total number of memory accesses: percentage over the total memory accesses that correspond to the dynamic memory manager

Once we had explored the key parameters of the memory manager and reduced the design space for the memory manager to a manageable size, we ran several simulations to analyze the performance of each option with respect to the number of memory accesses and memory footprint. A reduced design space allows performing an exhaustive analysis and obtaining the most suitable customized dynamic memory manager.

Table 3.5 Difference in memory accesses between each DMM and the optimum for each input

Input	DMM1 (%)	DMM2 (%)	DMM3 (%)	DMM4 (%)	DMM5 (%)	DMM6 (%)	DMM7 (%)	DMM8 (%)	DMM9 (%)
01	0.67	0.01	0.01	5024.59	4881.74	0	0	4986.61	4454.37
02	1.30	0.18	0.19	2240.55	2220.80	0	0.01	2268.15	2239.02
03	144.17	26.92	25.39	720.54	735.17	0	0.29	547.52	572.22
04	159.45	32.33	33.54	744.68	730.43	0	0.04	523.58	514.36
05	165.61	43.62	43.37	844.33	862.57	0	0.05	650.53	698.31
06	115.39	27.89	28.19	865.83	868.86	0	0.16	598.72	629.96
07	120.39	35.03	35.96	551.71	552.20	0	0.94	590.30	614.06
08	143.51	42.13	42.13	572.62	572.98	0	0	566.18	567.98
09	160.48	42.85	42.97	820.09	820.18	0	0.12	651.85	575.89
10	27.02	0.07	0.06	1620.66	1481.76	0.03	0	1420.79	1427.60
11	150.39	43.99	43.97	793.95	938.43	0.02	0	1080.67	931.31
12	142.78	37.94	38.03	834.40	835.10	0	0	806.23	962.48
13	66.95	19.41	19.48	1916.14	571.26	0	0.08	2863.90	1968.18
Average	107.55	27.10	27.17	1350.01	1236.27	**0**	**0.13**	1350.39	1242.75

DMM 6 and 7 cause the lowest overhead. The bolded data are the reference figures

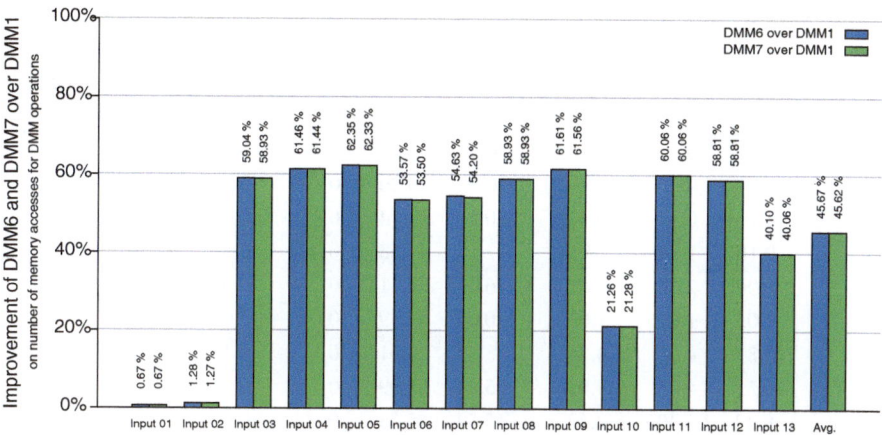

Fig. 3.8 Improvement on the number of memory accesses due to DMM when using DMM 6 or 7 instead of DMM 1

3.3.5.1 Reducing the Number of Memory Accesses

Having as target the minimization of memory accesses, the dynamic memory manager needs to be able to fetch blocks for the most popular sizes quickly. Similarly, the process of marking as free the blocks that become unused must be quick. This suggests the specialization of lists of free blocks for the popular sizes. However, using a high number of lists increases the search time, and therefore the number of lists must be reduced to the ones that receive most of the allocations.

Table 3.5 shows the difference in the number of memory accesses directly related to the management of dynamic memory to the optimal solution for each input case. DMMs 6 and 7 are the memory managers that incur the lowest overhead, with DMM 6 being the best one for most of the input cases (11 out of 13). Compared to DMM 1, the dynamic memory manager used as a reference for this case study, DMMs 6 and 7 obtain an improvement of up to 62.35 and 62.33 %, respectively (input set 5). On average, both managers enable a reduction of 45.67 and 45.62 % of the memory accesses due to the management of dynamic memory, respectively (Fig. 3.8).

Reducing the number of memory accesses due to DMM is important, but the final goal is to reduce the total number of memory accesses of the application. Table 3.6 shows the difference in the total number of memory accesses when using each one of the managers compared to the optimal for each input case. The results are consistent and, again, DMMs 6 and 7 enable the biggest reduction in memory accesses, with a negligible average deviation to the optimal for each input case. In a similar way with the previous analysis, Fig. 3.9 illustrates the improvement achieved by these two dynamic memory managers (up to 14.18 % for input 08 and 5.98 % on average).

As a way to understand the difference between the two solutions that are closest to the optimal for most cases, Fig. 3.10 shows the difference between them for each

Table 3.6 Difference in the total number of memory accesses caused by the dynamic memory managers and the optimum for each input

Input	DMM1 (%)	DMM2 (%)	DMM3 (%)	DMM4 (%)	DMM5 (%)	DMM6 (%)	DMM7 (%)	DMM8 (%)	DMM9 (%)
01	0.01	0	0	40.41	39.26	0	0	40.10	35.82
02	0.01	0	0	20.38	20.20	0	0	20.63	20.36
03	4.11	0.76	0.73	20.38	20.74	0	0.03	15.46	16.18
04	9.10	1.84	1.95	42.34	41.53	0	0	29.76	29.24
05	9.06	2.43	2.41	46.09	47.11	0	0.07	35.56	38.13
06	4.05	1.01	1.02	30.07	30.19	0	0.06	20.81	21.88
07	13.87	4.03	4.14	63.55	63.61	0	0.11	68.00	70.74
08	16.53	4.85	4.85	65.94	65.99	0	0	65.20	65.41
09	8.93	2.40	2.42	45.54	45.53	0	0.03	36.19	31.98
10	0.54	0	0	32.22	29.46	0	0	28.25	28.39
11	7.04	2.04	2.08	37.47	44.36	0	0.02	50.99	44.03
12	6.75	1.91	1.88	39.39	39.42	0	0.12	38.01	45.30
13	5.48	1.59	1.60	156.88	46.77	0	0.01	234.48	161.15
Average	6.57	1.76	1.78	49.28	41.09	**0**	**0.03**	52.57	46.82

DMM 6 and 7 minimize the number of memory accesses of the whole application. The bolded data are the reference figures

input case. The improvement of DMM 6 over DMM 7 is very small. This difference is further studied below.

3.3.5.2 Reducing the Memory Footprint

Using less memory is beneficial for an application running in an embedded system in many ways. For example, the system may use a smaller memory, which will translate into a faster access time and a lower energy consumption. The system manager may even be able to power down unused memory modules to further reduce energy consumption. Thus, reducing the memory footprint of the application is a relevant optimization goal.

In order to reduce the memory footprint of the application, the dynamic memory manager must ensure that the amount of memory wasted due to internal and/or external fragmentation is kept to a minimum. Therefore, an important optimization is to make sure that the blocks with the right sizes are used for each allocation. DMM 6 builds lists of free blocks for the most popular sizes (40, 1460 and 1500 bytes) to reduce the number of accesses needed to find the right block. DMM 7 adds some less frequently requested sizes (92 and 132 bytes) as well as intermediate ones (256, 512 and 1024 bytes) to avoid incurring a high penalty for allocations of intermediate sizes.

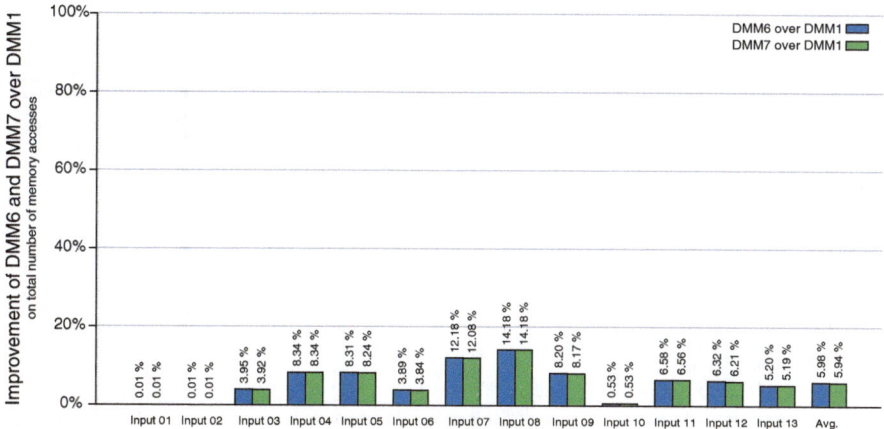

Fig. 3.9 Improvement on the total number of memory accesses due to the optimizations on the dynamic memory manager: improvement of DMM 6 and 7 over the reference memory manager (DMM 1)

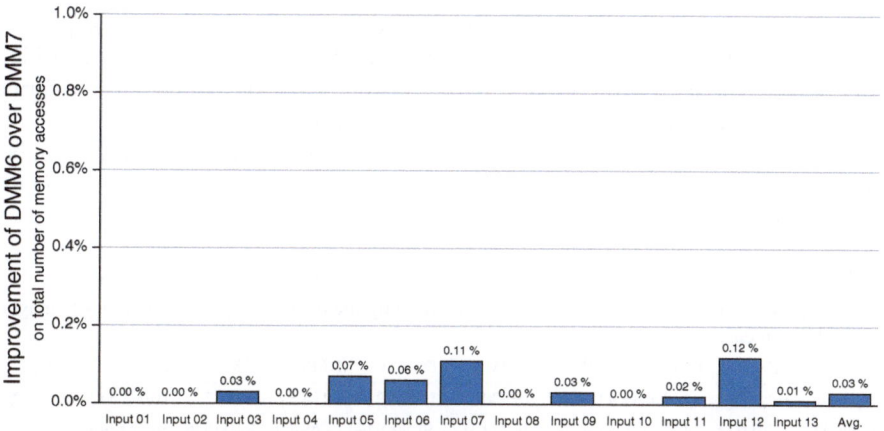

Fig. 3.10 Improvement on the total number of memory accesses due to the optimizations on the dynamic memory manager: marginal improvement of DMM 6 over DMM 7 for the total number of memory accesses

We performed several experiments to validate the previous assumptions based on the extracted software metadata. Table 3.7 and Fig. 3.11 show the amount of memory required by the application for each input dataset using each of the different dynamic memory managers. For the implementation of the DDTs, we used the previously selected combination: A3–B3. Table 3.8 presents the deviation of the different implementations with respect to the optimum for each input case. Considering the memory footprint, DMM 7 is the best memory manager for most of the input cases. However, the difference to a hypothetical perfect solution for all cases is now bigger,

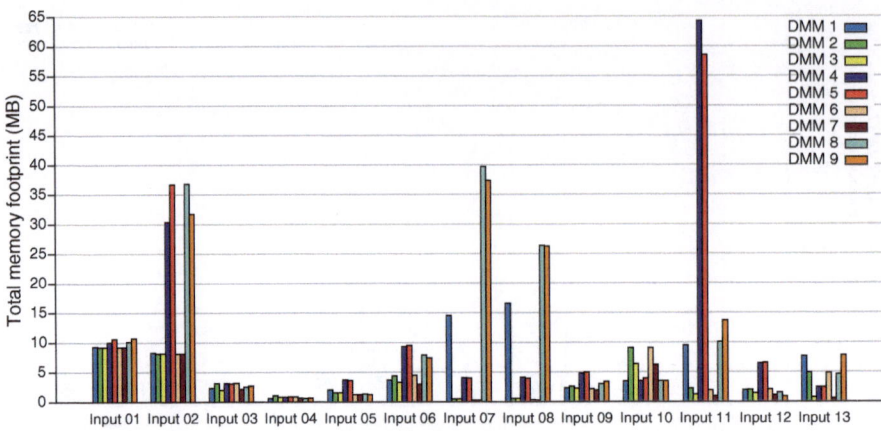

Fig. 3.11 Total memory footprint required by each of the dynamic memory managers for each input case

with an average deviation of 10.02 %, but a maximum error of 81.9 % (for input 10). On the contrary, the average difference of DMM 6 to the optimum is now considerably higher. The higher variation of the results obtained for the memory footprint, compared to the results obtained for the number of memory accesses, is due to the fact that the memory footprint is more sensitive to small variations in the number of blocks of each size that are allocated, even if the number of accesses to find the blocks remains the same.

Table 3.7 Total memory footprint (in bytes) required by the different dynamic memory managers for each input case

Input set	DMM1	DMM2	DMM3	DMM4	DMM5	DMM6	DMM7	DMM8	DMM9
Input 01	9 292 957	9 158 053	9 179 392	9 990 414	10 564 220	9 158 542	9 176 738	10 071 532	10 659 022
Input 02	8 261 672	8 114 640	8 146 305	30 344 312	36 688 944	8 069 856	8 100 836	36 775 346	31 719 557
Input 03	2 324 829	3 101 201	1 955 788	3 138 464	3 084 152	3 191 153	2 141 262	2 508 024	2 668 710
Input 04	583 720	1 068 413	793 154	805 870	848 280	858 652	638 234	564 347	618 625
Input 05	2 041 645	1 541 750	1 541 532	3 707 705	3 590 099	1 240 996	1 229 171	1 346 610	1 232 973
Input 06	3 657 349	4 387 276	3 302 356	9 287 451	9 497 449	4 450 461	2 957 479	7 873 662	7 400 638
Input 07	14 509 008	491 348	488 828	4 030 953	3 985 615	263 972	261 308	39 644 311	37 313 804
Input 08	16 552 416	535 652	510 296	4 077 566	3 946 176	279 673	249 007	26 375 050	26 260 057
Input 09	2 280 466	2 576 408	2 257 745	4 787 656	4 938 782	2 135 969	1 938 658	3 025 090	3 396 929
Input 10	3 441 933	9 070 597	6 375 234	3 487 045	3 944 477	9 064 424	6 260 861	3 511 306	3 510 688
Input 11	9 476 152	2 228 468	1 261 357	64 222 751	58 511 747	1 946 504	978 527	10 085 511	13 690 317
Input 12	1 959 322	2 038 829	1 417 485	6 459 893	6 600 368	2 093 518	1 122 728	1 581 100	897 078
Input 13	7 667 048	4 888 330	756 467	2 465 793	2 481 203	4 834 146	600 508	4 642 012	7 808 988

Table 3.8 Difference to the optimum for each input case of the total memory footprint required by the different dynamic memory managers. DMM 7 minimizes, on average, the difference to the optimum. The bolded data are the reference figures

Input	DMM1 (%)	DMM2 (%)	DMM3 (%)	DMM4 (%)	DMM5 (%)	DMM6 (%)	DMM7 (%)	DMM8 (%)	DMM9 (%)
01	1.5	0	0.2	9.1	15.3	0.0	0.2	10.0	16.4
02	2.4	0.6	0.9	276.0	354.6	0.0	0.4	355.7	293.1
03	18.9	58.6	0.0	60.5	57.7	63.2	9.5	28.2	36.4
04	3.4	89.3	40.5	42.8	50.3	52.1	13.1	0.0	9.6
05	66.1	25.4	25.4	201.6	192.1	1.0	0.0	9.5	0.3
06	23.7	48.3	11.7	214.0	221.1	50.5	0.0	166.2	150.2
07	5452.4	88.0	87.1	1442.6	1425.3	1.0	0.0	15071.5	14179.6
08	6547.4	115.1	104.9	1537.5	1484.8	12.3	0.0	10492.1	10445.9
09	17.6	32.9	16.5	147.0	154.8	10.2	0.0	56.0	75.2
10	0	163.5	85.2	1.3	14.6	163.3	81.9	2.0	2.0
11	868.4	127.7	28.9	6463.2	5879.6	98.9	0.0	930.7	1299.1
12	118.4	127.3	58.0	620.1	635.8	133.4	25.1	76.2	0.0
13	1176.8	714.0	26.0	310.6	313.2	705.0	0.0	673.0	1200.4
Average	1099.8	122.4	37.3	871.3	830.7	99.3	**10.0**	2144.0	2131.4

Figure 3.12 shows the difference in memory footprint of each dynamic memory manager to the optimum, averaged for all the input cases. Figure 3.13 shows an interesting measurement: the average overhead in memory footprint that each dynamic memory manager imposes. The application requires a certain amount of memory, but the dynamic memory manager needs some additional memory for its own internal structures. Additionally, the effects of internal and external fragmentation produce a significant increase in the actual amount of memory needed to serve the needs of the application. For the best suitable memory manager, DMM 7, the overhead is only 40.26 % (compared to 1 562.44 % of the reference manager, DMM 1 that is a general purpose memory manager). However, if the memory manager used is not fine tuned to the allocation behavior of the application, the overhead can increase up to 2 996.77 % (on average, when using DMM 8).

With the previous results, it seems logical to employ DMM 7 as the final memory manager for the application. However, in Sect. 3.3.6, the effect of both dynamic memory managers is analyzed independently when applying the optimizations on the transfer of blocks of dynamic data.

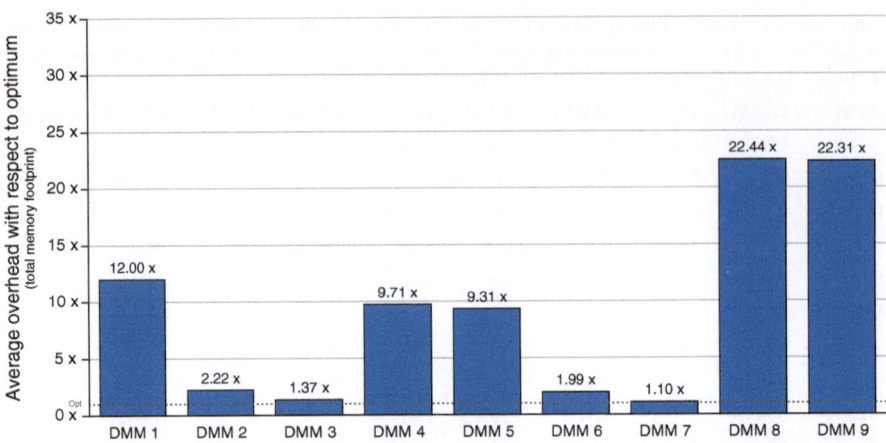

Fig. 3.12 Analysis of memory footprint with each dynamic memory manager: average difference in memory footprint of each memory manager to the optimum. Although DMM 7 is not the best solution for all the input cases, it is the best overall solution

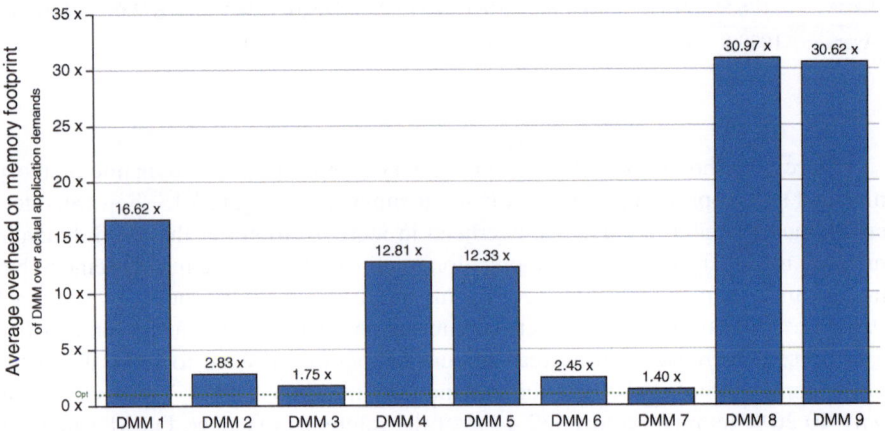

Fig. 3.13 Analysis of memory footprint with each dynamic memory manager: Average overhead in terms of memory footprint of each dynamic memory manager

3.3.6 Dynamic Memory Block Transfer Optimization

The traditional application of the DMA module is to transfer data in memory (or between memory and external devices), freeing up processor cycles. However, in embedded systems it is also common practice to use the DMA module to reduce the average latency to access data from the main memory (e.g., copying data to closer memories such as a "scratchpad" before the CPU actually needs them [50]). On the other hand, DRAM is usually the memory technology chosen to implement the main memory of embedded systems. These devices are internally organized in banks and pages, with the restriction that only one page can be active in each bank at any given time. Changing the active page in a DRAM bank has a non-negligible cost in terms of cycles and energy consumption. This type of organization favors sequential access patterns. However, when the DMA uses the main memory in parallel with the processor, accesses from both interleave in an undetermined way. Therefore, two relevant optimization goals for applications running on embedded systems are the efficient scheduling of data transfers for blocks of dynamic data using the DMA module and the right interleaving of accesses from the processor and the DMA to avoid unnecessary row activations in the pages of the DRAM modules. In this section, we evaluate the application of optimization techniques based on the information supplied by the software metadata for the transfer of dynamic data blocks.

The software metadata contains information on the number of block transfers that involve instances of dynamic data types, their length, direction (to/from main memory) and the thread that initiated them (mainly, the Block Transfers entity from Fig. 3.3). This information facilitates improving the utilization of the DMA module for the case study application of this example. If the input case produces long series of sequential accesses (i.e., the system processes mainly long packets), they are good candidates to be executed by the DMA. On the contrary, if the system has to process many small packets, the overhead of programming the DMA may be higher than the number of cycles that would be required had the transfer been performed directly by the processor. Additionally, the scheduling of accesses to dynamic data types must consider as well two more circumstances. First, when a block transfer is executed by the DMA in parallel with the processor, the external DRAM modules receive two simultaneous streams of accesses that may force extra row activations. Second, the DMA module can benefit itself from the lower latency of the DRAM burst modes; hence, it may be interesting to assure that the processor does not preempt the DMA module from the bus during transfers. However, it may be necessary to guarantee that the processor can access the memory in a bounded number of cycles, for instance, to fetch the code of an interrupt routine from the main memory.

Taking these considerations into account, we consider for this experiment three different scheduling policies. The first one executes all the accesses to dynamic memory with the processor. The second one uses the DMA module for blocks of more than 32 bytes (eight words), but the maximum number of cycles that the DMA engine may hold the bus during burst transactions is limited to eight words (therefore, once the DMA is granted access to the bus, it can transfer without interruptions at least as many bytes as the shortest transfer). Finally, the third configuration employs the DMA module for transfers of at least 32 bytes, but ensures that the DMA may access up to a full DRAM page in a single burst transaction to maximize the efficiency; additionally, this last policy uses the techniques presented in [135] and [22] to decide whether to use the DMA module or not if the system can recognize the current input case. We refer to these policies as "No DMA," "DMA Bad" and "DMA Opt", respectively. The results of this study reveal that the utilization of the DMA module can save a considerable amount of processor cycles (43 % on average when using the optimal DMA configuration with DMM 7), but only if the DMA is used appropriately.

Otherwise, it may have a significantly negative impact on the energy consumption of the memory subsystem, memory average latency and processor cycles wasted waiting to access the memory.

3.3.6.1 Analysis of the Improvements Achieved

Figure 3.14 shows the effect of a good scheduling: Using the right configuration, an improvement of up to 33 % in the number of cycles that the processor spends accessing memory is achieved in comparison with no using the DMA at all. There is also a small improvement in energy consumption.[6] Moreover, compared with a bad DMA configuration that does not limit sufficiently the interferences between the two elements, a 24 % of average improvement is possible for the number of DRAM row activations, 14 % for the number of processor cycles spent accessing memory and 9 % for the energy consumption.

Figure 3.15 illustrates the effect that the optimizations previously performed on the dynamic memory manager have on the performance of the system when using a

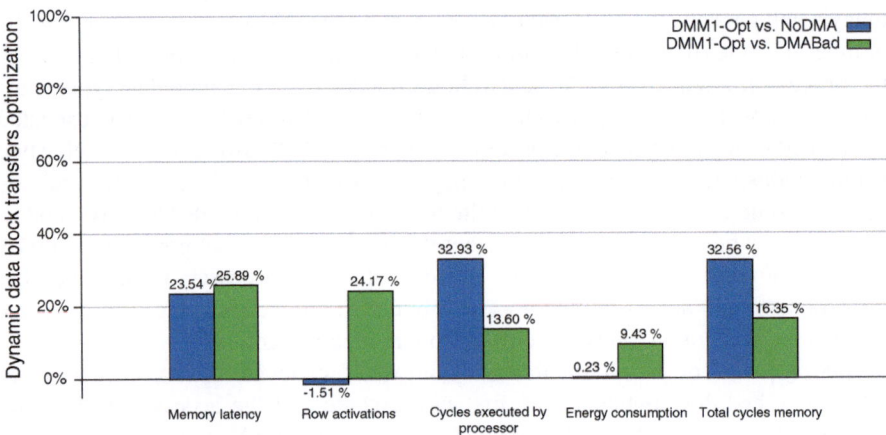

Fig. 3.14 Improvement achieved using a DMA module for block transfers of dynamic data. The *leftmost bars* show the improvements attained with the correct DMA configuration compared to not using the DMA. The *rightmost bars* show the improvements of using the correct DMA configuration, in comparison with using a wrong one that does not limit the interference between DMA and processor

---

[6] This improvement refers to the energy consumption in the memory subsystem. It is small because the processor and the DMA access the DRAM concurrently and, as the graph shows, there is a slight increase of row activations. The penalty in energy consumption of these additional row activations masks the benefits obtained by using the DMA. However, the total energy consumption of the system may be reduced much more because first, the DMA is more efficient accessing the memory than the processor and, second, the number of cycles that the processor spends accessing the memory is reduced, potentially allowing it to finish other tasks sooner.

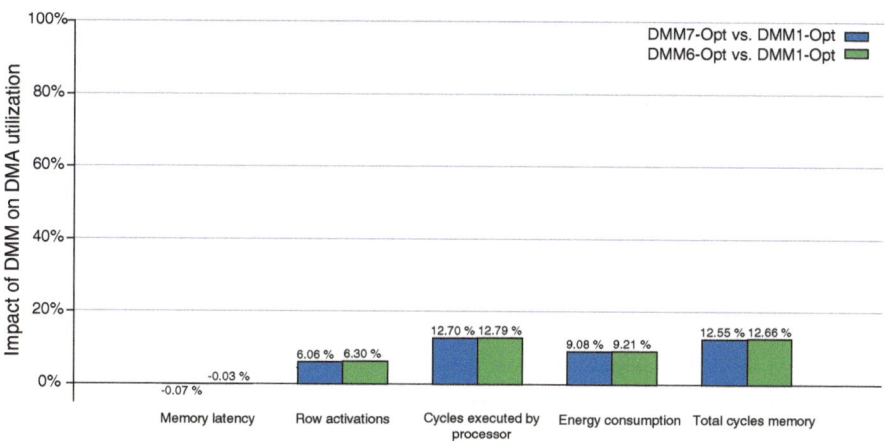

Fig. 3.15 Selecting one of the optimized dynamic memory managers improves the overall performance when adding a DMA module

DMA module. Here, we use the best DMA configuration with the dynamic memory managers selected in the previous step, DMM 6 and 7. Then, we compare their final performance results with the ones obtained using the same DMA configuration with the reference manager, DMM 1. The results of this comparison are revealing: The combination of the DMA with any of the optimized DMMs achieves improvements, in comparison with the combination of the DMA and the reference DMM, of about 6 % on the average number of DRAM row activations, 13 % on the number of cycles spent by the processor accessing the memory and 9 % in the energy consumption of the memory subsystem.

The outcome of the dynamic memory management optimization step was that DMM 7 is the best average solution when memory footprint is considered, and DMM 6 when considering the number of memory accesses. However, the difference between both when considering the number of memory accesses was negligible. Nevertheless, we kept around both dynamic memory managers just for the sake of analyzing their impact on the performance of the DMA module. Figure 3.16 shows that both managers have a very similar effect on the behavior of the whole memory subsystem, with the bigger impact being lower than 0.30 % (for the number of DRAM row activations). Therefore, we can now conclude safely that DMM 7 is the memory manager that should always be used in this system. The availability of a common software metadata representation simplifies this type of analysis.

Finally, Fig. 3.17 shows the overall improvements attained using DMM 7 with the optimal DMA configuration, so that the results of the DMMR and DMA optimization techniques are combined and jointly analyzed. Up to 43 % of the processor cycles are now free to be used for any purpose other than accessing dynamic data from the main memory, and a mean reduction of 9 % in the energy consumption of the

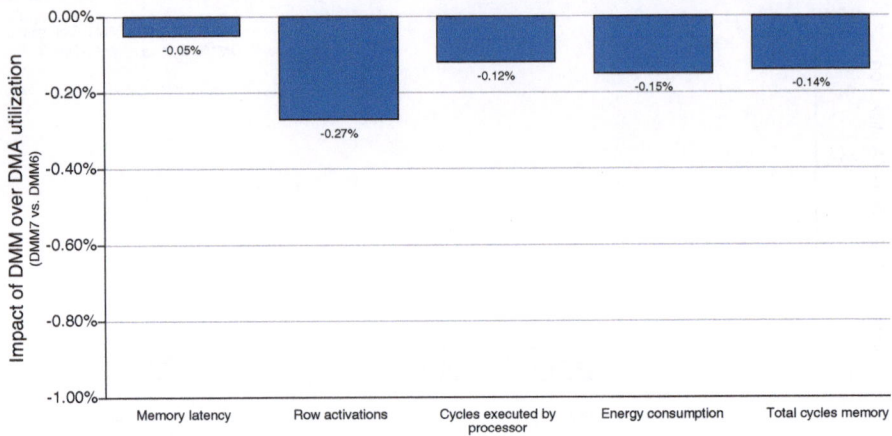

Fig. 3.16 Impact of the different optimized memory managers from the previous step on the utilization of the DMA: DMM 6 and 7 yield almost identical results

memory subsystem is possible. Moreover, compared to a wrong configuration of the DMA module, a 29 % reduction on the number of DRAM row activations and 18 % on the energy spent in the memory modules is possible.

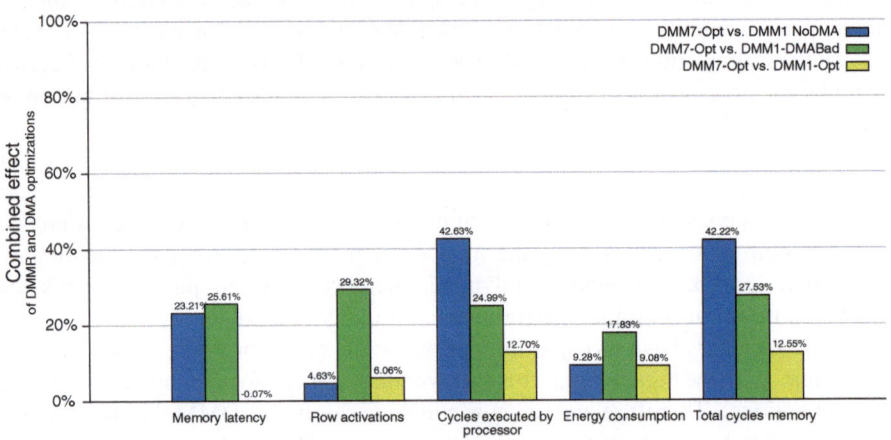

Fig. 3.17 Final combined effect of the DMM and DMA optimizations

3.4 Comparison to Related Work

The main differentiator in this chapter with respect to the related work is that our approach to metadata does not aim to define the engineering of the software applications, nor to analyze or characterize their structure. Instead, we try to represent the characteristics of their behavior, when they are subject to specific inputs, in a reusable way. Moreover, we specifically limit this work to the scope of applications dominated by dynamically allocated data types running on embedded systems. The produced software metadata can be used with the platform's hardware metadata to customize the resource management and apply different optimization techniques.

Regarding the subject of profiling, this chapter shows a method to annotate the application DDTs with templates so that all the accesses to their variables are logged. This method requires just one compilation of the application; it can then be run with different inputs to get different profiling runs. The main advantage of the explained annotation method is that the compiler propagates the logging instrumentation automatically, guaranteeing that all the accesses are logged. This method supports multithreading; however, as is the case with the logging of any multi-threaded code, it is possible that due to "Heisenbugs" the profiling alters lightly the execution trace of the application. We believe that this should not be a problem for well-behaved multimedia software that implements proper locking and thread-safety. In comparison, running the code with a tool such as Valgrind might help to keep the threaded execution of the code nearly identical to the execution without any profiling. However, it would also be harder to relate the profiling information to specific variables in the source code.

3.5 Conclusions

In this chapter we have introduced the concept of software metadata to characterize the dynamic data allocation and access behavior of software applications on embedded systems. We have defined both the concept of metadata and the relevant information needed to perform optimizations on the dynamic memory subsystem. This information can be shared and exploited by different optimization tools, even from independent vendors, enabling the optimization of complex embedded systems in terms of energy consumption, memory accesses and memory footprint. In fact, any tool can update metadata values for the metrics affected by its own optimizations. The updated information can then be used by subsequent optimization tools. The application of this methodology would reduce the overhead of characterizing the applications because the information generated during a single profiling and analysis phase would be available to all the subsequent optimization techniques.

The method to extract the metadata of embedded software applications that we have presented is divided in two steps: profiling of the applications, which produces raw information, and analysis of this information to obtain the final metadata values.

For the profiling step, we have proposed the use of a specialized template library. For the analysis step, we have shown a set of algorithms that can be used to infer different characteristics of the applications.

Finally, we have presented a case study in which software metadata allows to identify the most relevant characteristics of an application and enables the consistent utilization of multiple optimizations to achieve improvements in energy consumption, memory accesses and memory footprint.

Regarding future extensions, the concept of metadata can be exploited at run-time by enabling the dynamic selection of the DDTs and DMMs used to execute different phases (with different access patterns) of an application on a certain embedded system. This could be achieved through the use of system scenarios and a library based on combinations of simple elements to implement the designs produced by the DDTR and DMMR methodologies. The concept of system scenarios is based on the identification of relevant (different) input conditions at design-time and the computation of appropriate DDTR and DMMR solutions for each of these situations. Later, a monitoring module identifies at run-time the current scenario so that the operating system can implement the appropriate DDTR and DMMR solutions combining components from the library.

Chapter 4
Dynamic Data Types Optimization in Multimedia and Communication Applications

In future technologies of nomadic embedded systems an increasing amount of applications (e.g., 3D games, video-players) coming from the general-purpose domain need to be mapped onto an extremely compact device. Furthermore, the general trend of ubiquitous mobile access pushes developers to provide cross-platform applications with the same characteristics across a set of devices and desktop systems. Smartphones, tablets, in-car entertainment and navigation systems offer access to the same applications that traditionally run on PCs and servers. However, these new embedded systems, struggle to execute these complex applications because they come from desktop systems, holding very different restrictions regarding memory use features, and more concretely not concerned with an efficient use of the dynamic memory.

In fact, a desktop computer typically includes today between 4 and 8 GBs of RAM memory at least, as opposed to 512 MB or 1 GB present in modern embedded systems. Even as technology advancements lead to a convergence in specs, embedded systems have very different needs in memory and performance optimization. Providing instant power-on and power-off while preserving data integrity, strong connectivity, long battery life and comparable performance to desktop systems raises the need for an efficient port of complex dynamic applications on the dynamic memory subsystem. However, due to the increasing number of application types and their complexity, the mapping process is becoming more and more tedious, requiring a systematic approach.

Moreover, the available development time for application mapping has been dramatically decreased the last years, as new devices are introduced more and more often to the market. Engineers usually do not have enough time to optimize the applications so that they take full advantage of the memory subsystem. Instead, based on experience and analysis tools they focus on the most pressing issues, which unfortunately can lead to the system being operated in a local optimum, loosing a potentially better global optimization point. Thus, there is an increasingly pressing need for a high level systematic (and at the same time automatic approach) for the mapping of complex dynamic applications onto embedded systems handling multimedia and network applications. As a matter of fact, one of the main tasks of the mapping process of such applications is the optimization of the dynamic memory subsystem.

© Springer International Publishing Switzerland 2015
D. Atienza Alonso et al., *Dynamic Memory Management for Embedded Systems*,
DOI 10.1007/978-3-319-10572-7_4

In modern dynamic applications, data is stored in entities called data structures, *Dynamic Data Types* (DDTs) or simply containers, like arrays, lists or trees, which can adapt dynamically to the amount of memory used by each application [173]. These containers are realized at the software architecture level [99] and are responsible for keeping and organizing the data in the memory and also servicing the application's requests at run-time. These services require to include abstract data type operators for storing, retrieving or altering any data value, which is not tied to a specific container implementation and shares a common interface (as in STL [145]). The implementation of these operators depends on the chosen container and each one of them can favor specific data access and storage patterns. Each application can host a number of different containers according to its particular data access and storage pattern in the algorithm. Choosing an improper operator and container implementation for an abstract data type will have significant negative impact on the dynamic memory subsystem of the embedded system [21]. On the one hand, inefficient data access and storage operations cause performance issues, due to the added computational overhead of the internal DDT mechanisms. On the other hand, each access of the DDTs to the physical memory (where the data is stored) consumes energy and unnecessary accesses can comprise a very significant portion of the overall power of the system [21, 52]. In fact, energy consumption is one of the most critical limiting factor in the amount of functionality that can be placed in these devices, because portable computers like tablets and notebooks running complex multimedia and network applications rely on limited battery energy for their operation. As a result, the optimization of the DDT for a certain nomadic embedded system needs to consider four main factors:

1. The access pattern over time of the algorithm implemented (temporal locality). If some (dynamic) data is reused throughout the entire application, e.g., in the form of a small dynamic buffer, it will occupy precious internal memory, and will affect the available memory space for other dynamic data to be stored in memory levels close to the processor.
2. The amount of memory accesses. If the data is present in the processor registers, the elements have no access penalty. Since the number of registers is limited, access will be needed from lower, larger and slower levels of the memory hierarchy and each access to such a lower memory hierarchy can result in CPU stalls if this is not properly addressed.
3. The size of the required data. Just as in statically dominated applications, the size of the data has an important impact on the use of the memory hierarchy. Contrary to the static data types, the DDT can be refined to exploit this memory hierarchy.
4. The mechanisms to access the data (as defined by the data structures of the system). The retrieval of a particular data element can incur several other accesses in complex data types. For example, accessing a random element in a double linked list (an often used dynamic data type) requires on average $\frac{N}{4}$ pointer dereferences.

Taking all these previous characteristics into account, it is clear that a suitable design of the DDTs need to be developed at the system-level in order to handle the complexity of new nomadic embedded systems. On the one hand, within this exploration and optimization process of dynamically allocated data within the source code of a certain application, the first issue is the definition of the design space of DDT implementations. In fact, the designer must choose the best among a large number of possible DDT implementations [8, 173] (dynamic arrays, linked lists, etc.), according to the specific restrictions of typical embedded design metrics, such as, performance, memory footprint and energy consumption. This task is typically performed using a pseudo-exhaustive evaluation of the design space of DDT implementations (i.e., multiple executions) for the application to attain the Pareto front [52]. That search would try to cover all the possible optimal implementation points for the aforementioned required design metrics. This exhaustive construction of the Pareto front is a very time-consuming, sometimes even unaffordable, process. Moreover, due to the inter-dependencies between DDTs, namely, that one DDT implementation behavior may affect the performance or memory footprint of another one [52], the refinement process must explore the whole range of possible combinations of the different DDTs. Thus, the number of experiments to be carried out typically becomes unaffordable even for a small number of DDTs. For instance, in the case of an embedded application including 9 different DDTs that need to be explored for 10 basic relevant implementations of DDTs for multimedia applications (as proposed in [13, 21, 100]), the number of experiments (i.e., multiple runs of the application) that need to be performed is 10^9; testing all these combinations manually is not feasible. In addition, existing optimization techniques for embedded systems [174] relying on partial cost estimators to enable covering and pruning of the design space of possible DDT implementations have been proposed, the exploration phase for complex applications still takes days. This is due to the lack of methods to capture the aforementioned inter-dependencies and collateral effects of multiple DDTs interacting together, typical in the latest dynamic multimedia applications ported to nomadic embedded systems (e.g., games or scalable video rendering applications [52, 152, 153]). Hence, the optimization of DDT implementations for nomadic embedded systems requires new exploration and optimization methods with respect to more traditional approaches targeting embedded systems relying mainly in static data allocation.

On the other hand, a major issue (from the coding and verification of the final source code viewpoint) is extending the set of available DDTs with new implementations of multi-layered (complex) DDTs to target the access requirements of new application domains ported to nomadic embedded systems (e.g., communication applications). When DDTs are used in programs, they can be implemented by the developer in the most naive form (because the developer is more focused on the implemented algorithm within the application) or in a manually optimized implementation where the number of implementation alternatives is defined by the experience and inspiration of the developer. The former case often results in an unacceptable power and performance reported figures of the ported application for the final embedded system [52], while the latter has proven to be programming intensive [13, 52]. Even when standardized languages (if used at all) offer considerable support, the developer still has to define the access pattern on a case per case basis, as the target nomadic embedded platforms are often not equipped with extensive hardware and system support for dynamic memory. Therefore, new programming methods that can be used to build and refine complex layered DDTs from basic ones in a modular way are needed. As a result, early design-flow estimates on implementation trade-offs to refine the system design can be obtained.

In this chapter we present a methodology that addresses the aforementioned problems related to the design and optimization of DDTs for dynamic multimedia and network applications targeting nomadic embedded systems, both for software and system integration engineers. We do not cover here the optimization of statically allocated data structures, which are covered in detail in other books in the literature [35], we refer the interested reader to those publications for complete overviews on static memory management. On the one hand, the proposed design and optimization methodology provides software engineers with the necessary interface of access methods and DDT implementations to cover applications using either pure imperative languages (e.g., C code) or object-oriented languages (e.g., C++) using the concept of data containers. As a result, this methodology allows software designers to reach a complete abstraction of the underlying DDT implementations and their related implementation problems (e.g., pointer management, definition of interfaces or access methods, etc.). On the other hand, the presented methodology is also valid for system integration engineers because it simplifies the complex process of porting applications originally designed for a different target platform. In this regard, we present a novel, semi-automated, optimization approach for the DDTs of multimedia and network applications, which can target either a concrete user-defined multi-objective optimization metric (i.e., memory footprint, memory accesses, energy consumption or linear combinations of them), or a number of overall solutions that respect the defined user constraints (Pareto front). In a dynamic environment where a nomadic embedded system has to operate in different working points throughout the execution of the application, these Pareto fronts are crucial. Otherwise, the conventional solution with only a single working point would become highly suboptimal.

This chapter is organized in the following way. In Sect. 4.1, we review the existing work related to the optimization of dynamically allocated data structures (or DDTs) for high-performance embedded systems. Next, in Sect. 4.2, the dynamic behavior of multimedia and network applications domains are analyzed, which characterizes the conditions for our design methodology to be applicable. Section 4.4 then continues with a case study of the characterisation approach. According to this analysis, in Sect. 4.4 we present a set of systematic, high-level, data-structure transformations that enable the optimization of DDTs according to the specific design constraints of the target nomadic embedded system. Then, Sect. 4.5 illustrates the application of the DDT transformations to a specific application, providing insights on the effects of each transformation. Next, in Sect. 4.6 we present the complete multi-objective exploration and optimization methodology of DDT implementations, which relies on the definition of a complete set of complex multi-layer DDT implementations according to the previously presented transformations and an analytical pre-characterization of the possible possible elementary DDT blocks. In this case, we will distinguish between the set of complex DDT implementations used in the context of plain imperative languages, such as C, or object-oriented ones, e.g., C++. Then, in Sect. 4.7, we illustrate the application of the methodology to several case studies of multimedia and network applications of latest nomadic embedded applications. Finally, in Sect. 4.8, we summarize the main conclusions of this chapter.

4.1 Related Work

To optimize the use of dynamic memory, the designer must choose among a number of possible DDT implementations [8, 173] (dynamic arrays, linked lists, etc.) the best one in each case, according to the specific restrictions of typical embedded design metrics, such as, performance, memory footprint and energy consumption. This task is typically performed using a pseudo-exhaustive evaluation of the design space of DDT implementations (i.e., multiple executions) for the application to attain the Pareto's front [52], which would try to cover all the optimal implementation points for the aforementioned required design metrics. The construction of this Pareto's front is a very time-consuming process, sometimes even unaffordable without proper DDT refinements and exploration methods. For instance, in the case of an embedded application including 9 different DDTs that need to be explored for 10 basic relevant implementations of DDTs for multimedia applications (as proposed in [13, 21, 100]), the number of experiments (i.e., multiple runs of the application) that need to be performed is 9^{10}, which is not feasible to be manually tested.

Regarding DDT refinement, in general-purpose software and algorithms design [8, 173], primitive data structures are commonly implemented as mapping tables. They are used to achieve software implementations with high performance or with low memory footprint. Additionally, the Standard Template C++ Library (STL) [145] or other proposed templates [33] provide many basic data structures to help designers to develop new algorithms without being worried about complex DDT implementation issues. These libraries usually provide interfaces to simple DDT implementations and the construction of complex ones is a responsibility of the developer. Furthermore, these libraries focus exclusively on performance and while they can be considered as acceptable general-purpose solutions, they are not suitable for new generation embedded devices, where performance, energy consumption and memory footprint must be optimized together.

For embedded software, suitable access methods, power-aware DDT transformations and pruning strategies based on heuristics have started to be proposed for multimedia and network applications [52, 174]. However, these approaches require the development of efficient pruning function costs and fully manual optimizations. Otherwise they are not able to capture the evaluation of inter-dependencies of multiple DDTs implementations operating together. In addition, a very recently proposed method to explore the DDT implementation design space shows the application of evolutionary computation [16]. This method relies on the definition and the analytical pre-characterization of the possible elementary DDT blocks, which are subsequently used in a *Genetic Algorithm* (GA) of type *Vector Evaluated Genetic Algorithm* [143]) to model the existing inter-dependencies of using different DDTs implementations.

Finally, this modeling of inter-dependencies is used to prune the design space, and to select the best choice according to the user's metrics with quite successful results in multi-objective exploration constraints. Also, several transformations of data structures for compilers have simplified local loops in embedded programs [126]. Nevertheless, they are not suitable for exploration of complex DDTs employed in modern multimedia and network applications, because they handle only very simple data structures (e.g., arrays or pointer arrays), and mostly focus on performance.

Also, new methods for a modular construction of custom high-level components in DDT implementation can be envisaged with *abstract derived classes* or mixins [150]. This programming technique has been inherited by functional programming languages (e.g., Lisp) and has already been used for quite some time in large reusable modules of object-oriented design. Very recently, it has been shown how to apply this programming technique to describe a high-level approach for modeling and refining complex DDTs using abstract derived classes in C++ [15]. This approach enables the multimedia developer to compose, evaluate and refine complex data types in a conceptually straightforward way, without a time-consuming programming effort.

In addition, according to the characteristics of certain parts of multimedia and network applications, several transformations for DDTs and design methodologies [23, 35, 50] have been proposed for static data profiling and optimization considering static memory access patterns to physical memories. In this context, the use of GA-based optimization has been applied to solve linear and non-linear problems by exploring all regions of the state space in parallel. Thus, it is possible to perform optimizations in non-convex regular functions, and also to select the order of algorithmic transformations in concrete types of source codes [118, 131]. However, such techniques are not applicable in DDT implementations, due to the initially unpredictable nature of the data to be stored at compile-time.

Finally, compiler techniques for code compaction and minimization of energy and power consumption in high-performance systems are also partially applicable in parts of the source code of DDT implementations that can be analyzed and tuned statically (e.g., number of registers for loop traversals in DDT solutions, etc.) [23, 56].

4.2 Analysis and Characterization of Multimedia and Wireless Network Applications

The optimizations in the proposed methodology are effective because they take profit of the characteristics of embedded multimedia and network applications use DDTs with a set of common features. These features are described in this section, and have been obtained after an exhaustive analysis of a representative set of industrial applications in these research domains. Among this set of analyzed multimedia and network applications, there are 3D image reconstruction applications [45, 138], video rendering applications as the MPEG-4 *Visual Texture Coder* (VTC) [59, 124], 3D games [71, 122], the URL-based context switching application [116], IPv4 routing algorithms [91] and firewall applications [116].

4.2.1 Application Specific Access Behavior

In all these previous types of applications it is possible to distinguish between three typical access patterns: sequential access and random accesses. In particular, sequential accesses are predominant in multimedia and network applications and imply three different types of operations, namely, insertion, removal, and traversal. In addition to this, we can realistically assume that when a record is inserted it is not already present in the data structure. Similarly, when a record is removed, it is indeed the case that the record is a priori stored in the data structure. All these characteristics are typical in the multimedia and network applications, and in the proposed DDT optimization methodology we exploit them to define a complete set data structure transformations to optimize the memory footprint, memory accesses and overall power consumption of the underlying DDT implementations. The proposed data-structure transformations themselves are, however, invariant to these characteristics. Therefore, they are applicable on other case studies and additional application domains with similar behavior.

4.2.2 Representative Sizes and Types of Basic Allocated Elements

In multimedia and network applications for nomadic embedded systems, the range of possible dynamic applications is in theory initially very broad. However, for most of the embedded systems the number and types of multimedia and network applications to be included in the final design (at least to a large extend) are known at design-time. Thus, it is feasible to analyze the types of dynamically-allocated objects that exist in each of them (e.g., points, triangles, 3D faces, acknowledgement or frequent packets sizes, etc.) and to design the most convenient DDT implementation for each variable. Furthermore, our experience has shown that, in each application, the range of sizes used by the dynamic elements is very limited and can be determined at design time by, first, a static analysis of the source code of the application and, second, a profiling analysis of the application under exploration with a reduced number of input data sets (i.e., in general no more than 10 variations). Note that the size of the dynamic elements is known a priori, but not the actual number to allocate, which can vary significantly from one possible representative input to another; thus, the use of dynamically allocated memory is justified and extensively used in these application domains. For example, in order to analyze an MPEG-4 video player, different system configurations, such as screen resolution or visualization size, need to be studied according to the target embedded device: a *Portable Digital Assistant* (tablet), a mobile phone, a video-game station, etc. This set of configurations needs to be explored for a representative set of input frames (e.g., with a different number of rendered objects and textures, etc.), while taking into account the probability distribution of the different types of inputs for the final working conditions of the target embedded device, as this features will influence the final dynamic data structure to be used for the final application.

In addition, it is needed to recognize the dominant dynamically-allocated elements of each application. In fact, each application can contain a large number of allocated elements, but in most of the multimedia and network applications few variables (i.e., between 5 and 20) tend to account for a very large percentage of the overall memory accesses and/or memory footprint used for dynamic memory storage, in general between 50 and 70 %.

4.3 Example: 3D Image Reconstruction System

The previously outline characteristics are illustrated using a new 3D image reconstruction application [160]. In particular, we focus on one of the internal algorithms that works like 3D perception in living beings, where the relative displacement between several 2D projections is used to reconstruct the 3rd dimension [138]. This software module heavily uses dynamic memory and is one of the basic building blocks in many current 3D vision algorithms: *feature selection and matching*, which involve multiple sequential accesses and input data filtering operations. The algorithm studied has been extracted from the original code of the 3D image reconstruction system (see [160] for the full code of the algorithm with 1.75 million lines of high level C++), and creates the mathematical abstraction from the related frames that is used in the global algorithm. The algorithm selects and matches features (corners) on different subsequent frames and the relative offsets of these features define their spatial location (see Fig. 4.1). The operations done on the images are particularly memory intensive, e.g., each image with a resolution of 640 × 480 uses over 1 Mb, and the accesses of the algorithm to the images are randomized. Thus, classic image access optimizations as row-dominated accesses versus column-wise accesses are not relevant.

The number of generated candidate matches is highly dependent on a number of factors. Firstly, the *image properties* affect the generation of the matching candidates. Images with highly irregular or repetitive areas will generate a large number of localized candidates, while a low number will be detected in other parts of the image. Secondly, the *corner detection parameters* have a profound influence on the results (and, consequently, on the input to the subsequent matching algorithm) because they affect the sensitivity of the algorithm used to identify the interesting features in the images/frames [72]. Finally, the *corner matching parameters* that determine the matching phase have a profound influence and are changed at run-time (e.g., the acceptance and rejection criterion changes over time as more 3D information is retrieved from the scene).

Taking all the previous factors into account, the possible combinations of parameters in the system make an accurate estimation of the memory cost, memory accesses and energy dissipation at compile time very hard or nearly impossible. This unpredictable memory behavior can be observed in many state-of-the-art 3D vision

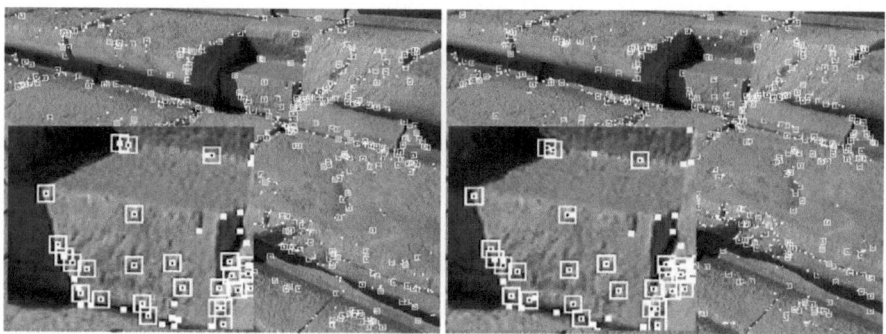

Fig. 4.1 Initialisation of the matching of corners on two images of the steps of an amphitheater (archaeological site): based on neighborhood search. Already most matches seem to be correct (partially due to the minor difference between the images, which can be seen at the *right hand bottom corner*). Part of the center is enlarged

algorithms and makes them very difficult to optimize with traditional compile-time optimization methodologies for embedded systems. Nevertheless, since this metric 3D-reconstruction from video algorithm can perform the reconstruction of 3D scenes from images and requires no other information apart from multiple frames, it is especially useful for situations where extensive 3D setup with sensitive equipment is extremely difficult, e.g., crowded streets or remote locations, or impossible as when the scene is no longer available [45, 60]. Hence, it is extensively used for quick on-site visualization, and speeding up the application to be able to process more frames for a more detailed reconstruction is a very important problem, which demands extensive code transformations and optimizations of the dynamic memory subsystem. Also, energy consumption is paramount for hand-held visualization devices and needs to be optimized as well. In the following paragraphs the internal DDTs of this case study are explained and their respective dynamic allocation behaviors are carefully analyzed.

The algorithm uses internally several DDTs whose sizes cannot be determined until the system is running, because they depend on factors (e.g., textures in the images) determined outside the algorithm (and uncertain at compile-time). Furthermore, due to the image-dependency related data, the initial DDT implementations for the variables do not fit in the internal memory of current embedded processors. These DDTs are the following:

- `ImageMatches` (`ImageMatches`) is the list of pairs where one point in the first image, matches another one on the second image based on a neighborhood test [138].
- `CandidateMatches` (`CandidateMatches`) is the list of candidates that must go through a normalized cross-correlation evaluation of a window around the points [160].
- `MultImageMatches` (`MultiMatches`) is the list of pairs that pass the aforementioned evaluation criterion. Still one point from the first image can be listed in multiple candidate pairs for a later best match selection.

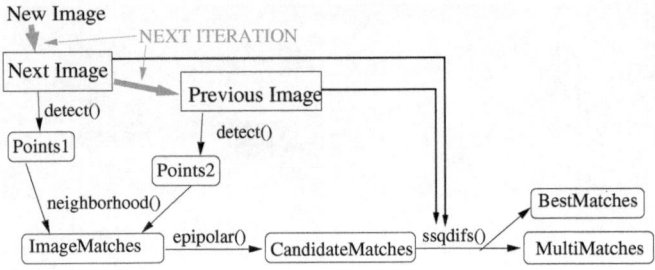

Fig. 4.2 Execution flow of the DDTs in the 3D image reconstruction application

- BestMatches (BestMatches) is the subset of pairs where only the best match for a point (according to the test already mentioned) is retained.

These DDTs were originally implemented using double linked lists (see Sect. 4.6.2 for more details) and exhibit an unpredictable memory size, typical in many state-of-the-art 3D vision systems [160] since they use some sort of *dynamic candidate selection* by traversing the input data in a sequential way, followed by *a criterion evaluation*. Figure 4.2 shows this behavior in the generation order of the DDTs. First, ImageMatches is generated after a neighborhood test is applied to pairs of corners detected in two images. Then, CandidateMatches is created using the information acquired from previous images. Finally, MultiMatches and BestMatches are generated with the pairs that pass a normalized cross correlation evaluation [138].

Although the global interaction between these DDTs is defined by the algorithm (see Fig. 4.2), it is not clear how each DDT affects the final system figures (e.g., memory footprint) until the application is profiled. Therefore, our methodology is used to explore the behavior of the different DDTs. After inserting our profiling library and running the tools as presented in this book in Chap. 3, profiling information is obtained. Then, a memory use graph is automatically generated (Fig. 4.3) and memory accesses, memory footprint and energy dissipation figures are calculated (see Table 2.1). In order to account for the varying memory use during the program run, normalized memory is used to get an estimation of the overall contribution of each DDT to the energy dissipation shown in Table 2.1). This attributes more accurate contributions to energy cost estimates and avoid that, for instance, DDTs with very short and high memory usage, distort and hide the memory contribution from other DDTs. For the energy estimations, the embedded SRAMs memory model described in [84] is used, considering a $0.13\,\mu$ technology node for the memory hierarchy implementation. This model depends on memory footprint factors (e.g., size, internal structure or leaks) and factors originated by memory accesses (e.g., number of accesses or technology node used). Note that any other model can be used just by replacing that module in the tools.

The analysis of the profiling information (Table 2.1 and Fig. 4.23) shows how the DDTs affect the dynamic memory subsystem. The DDTs and their different dynamic behaviors can be derived from the internal subdivision of the global algorithm in

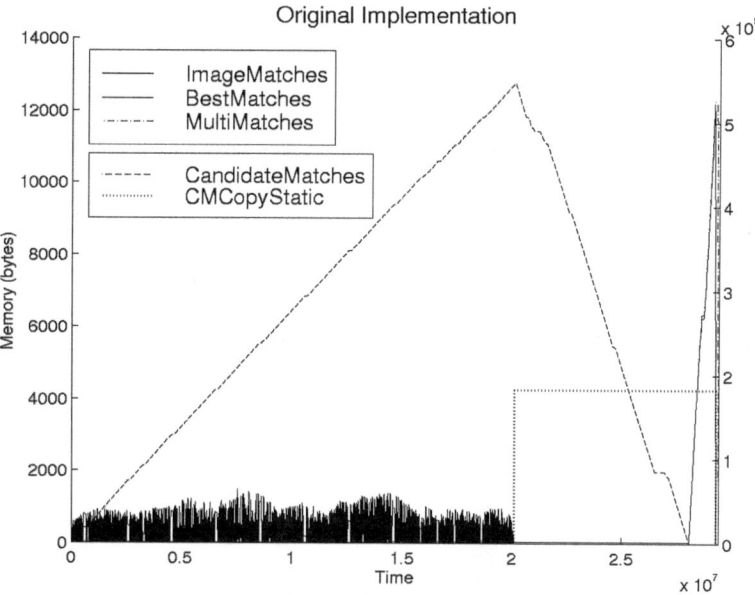

Fig. 4.3 Memory footprint utilization over time in the original implementation of the 3D image reconstruction case study. All plots are mapped on the left axis, except CandidateMatches and CMCopyStatic (right axis)

different sub-algorithms previously described, which include different traversal and filtering phases. First, CandidateMatches is the largest DDT in this system. Secondly, ImageMatches has frequent accesses. Finally, CMCopyStatic is a dynamic array implementation, which has a much faster access to specific stored data, that only keeps a copy of the content of CandidateMatches) consumes an important part of the energy used by the system. As this analysis shown, the DDTs in multimedia applications interact as local variables (with very limited life-time) that store the intermediate data generated between the different sub-algorithms that conform the global algorithm (e.g., ImageMatches or CMCopyStatic), or as variables that store the filtered input data between the different sub-algorithms, and which are used in the end to provide the output data of the current module to the next software modules of the overall applications, due to modularity in software engineering, as it is the role of BestMatches and MultiMatches. A final characteristic of multimedia and network applications is the repeating cycle of their dynamic allocation and de-allocation patterns. In particular, in the considered 3D image reconstruction module, this previously analyzed dynamic behavior is similar for each pair of consecutive frames, as in each new iteration of the algorithm, the oldest frame is replaced by the next one in the input set of frames and the algorithms are applied in the same way (Fig. 4.4).

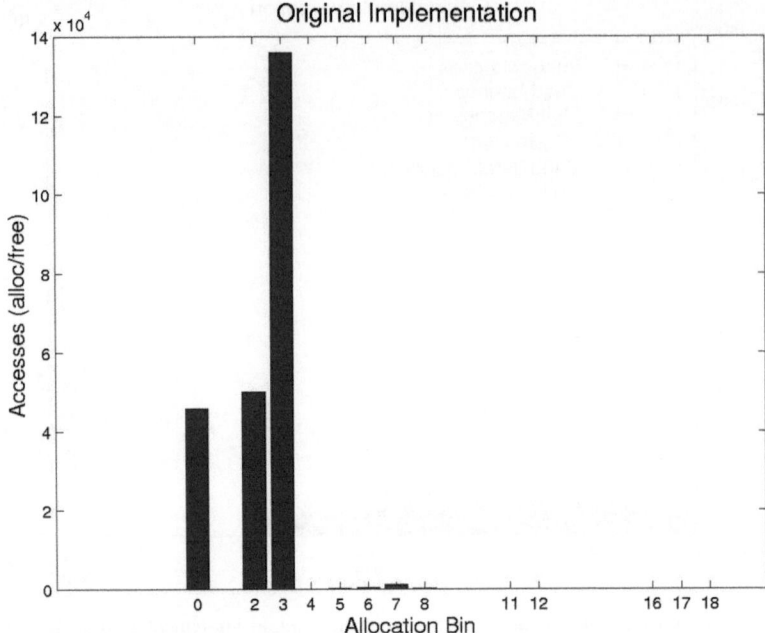

Fig. 4.4 Study of block-size allocation bins of the Kingsley DM manager [171] requested by the original 3D reconstruction application

4.4 Transformations of Dynamic Data Types

After the analysis of the common characteristics of the DDTs included in multimedia and network applications, in order to define a complete design and exploration methodology of DDTs, we must define a set of transformations and refinements of DDT implementations suitable for these two domains of applications. The definition of this set of transformations and refinements of DDTs need to be manually validated on a representative set of applications of the explored domains, while they are executed on a complete set of target nomadic embedded architectures.

Once this representative set of transformations has been defined for the considered domains of applications, it is possible to design a complete library of DDT implementations that will enable an automatic exploration of optimal DDT implementations, as presented in Sect. 4.6.

As a result, in the following subsections we present the set of data structure transformations covering the multimedia and network application domains, and intuitively motivate the usefulness of each transformation in terms of memory footprint and/or data accesses for specific examples.

4.4.1 Adding a Linked Structure

A simple example of high-level data structure transformation is presented in Fig. 4.5. In Fig. 4.5, a data structure consisting of one array (i.e., data structure 1) is transformed into a data structure consisting of two arrays and a head pointer (i.e., data structure 2), by adding links to data structure 1. The head pointer is a logical pointer or link to the first entry of the data structure. It merely contains the index (and not the physical address) of the array element to which it is "pointing" to.

Data structure 1 is a sparse array (AR) of records: it contains fragmentation. Each record is denoted by 'R'. The index of the corresponding array element is denoted by 'K'. This index value is equivalent to an implicit key, i.e., the index or key value is not stored explicitly. Every record R corresponds to one unique position in the array. This for instance means that record R2 can only be stored in the array element that is characterized by K2 and nowhere else. Data structure 2 is the transformed data structure: a linked layer (i.e., the top array with the head pointer) is introduced.

The top array of data structure 2 contains links which are merely index values (or logical pointers). Every array element of both the first and the second array (of data structure 2) corresponds to a unique record R. In other words, fragmentation is present in both arrays. This results in a larger memory footprint for data structure 2 in comparison to data structure 1.

An implicit correlation is present between the array elements of the top array and the array elements of the bottom array of data structure 2. For instance, if the second array element of the top array is consulted, then the associated record (i.e., R2) in the bottom array can be consulted directly as well with a second data access (i.e., no overhead accesses are present).

Implicit correlations are not depicted explicitly in Fig. 4.5 and other figures. Intuitively, finding a specific record is as cheap for data structure 2 as it is for data structure 1: just one data access is needed. The link layer is not used for this operation. Traversing all records is cheaper for data structure 2 than data structure 1 due to the links that can be used to quickly traverse the whole data structure.

To summarize, transforming data structure 1 without a linked layer into data structure 2 with a linked layer has the disadvantage that memory footprint increases. However, the data accesses needed to traverse all records decrease. Hence, trade-offs exist between memory footprint and memory accesses required for a complete traversal of the stored data in the DDT when this transformation is used. The number of data accesses needed to find a specific record remains unchanged.

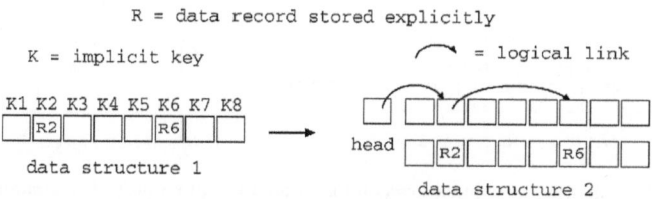

Fig. 4.5 Adding an array of links on top of an array of records

4.4.2 Implicit Versus Explicit Keys

In Fig. 4.6, a data structure consisting of one array (i.e., data structure 1) is transformed into a data structure consisting of two arrays and a tail pointer (i.e., data structure 3). The tail pointer is a logical pointer or link to the last entry of the data structure.

Data structure 1 is the initial sparse array of records. The top array, of data structure 3, contains the key values of the array of data structure 1. In other words, the keys are stored explicitly. The bottom array, of data structure 3, is similar to the array of data structure 1. In addition to this, every record R can potentially be placed any where in this bottom array.

An implicit correlation is present between the array elements of the top array and the array elements of the bottom array of data structure 3. For instance, if the second array element of the top array (i.e., K6) is consulted, then the associated record (i.e., R6) in the bottom array can be consulted in a second data access without additional accesses.

The transformation presented in Fig. 4.6 results in a non fragmented representation (cfr., data structure 3). We assume in this specific example that it is a priori known that maximum four (out of eight) records are to be stored in the data structure at any moment in time. Based on Fig. 4.6, data structure 3 consumes slightly more memory than data structure 1.

Based intuitively on Fig. 4.6, finding a specific record is more costly, in terms of data accesses, for data structure 3 than it is for data structure 1. This is because in data structure 3 the record to be found can be stored any where in the data structure. On the other hand, traversing all records is cheaper, in terms of data accesses, for data structure 3 than it is for data structure 1. Five data accesses are needed for data structure 3 while eight are needed for data structure 1 (since it is not known which two records are stored in the array).

To summarize, transforming data structure 1 with implicit keys into data structure 3 with explicit keys has the advantage that traversal becomes less expensive in terms of data accesses. On the other hand, the memory footprint and the data accesses needed to find a specific record increase. However, in some cases, the traversal operation can be simplified to only traversing the records (R) and not the explicit keys (K) since it is not necessary to know the key values in that particular context. This reduces the number of data accesses for traversal.

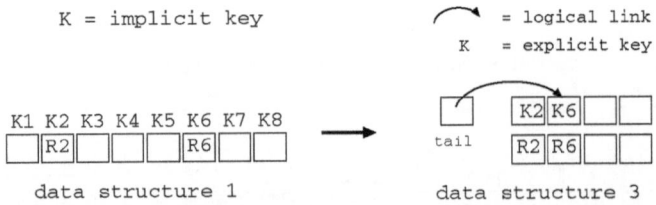

Fig. 4.6 Adding an array of explicit keys on top of an array of records. The assumption is made that maximum four records are stored in the data structure (at any moment in time)

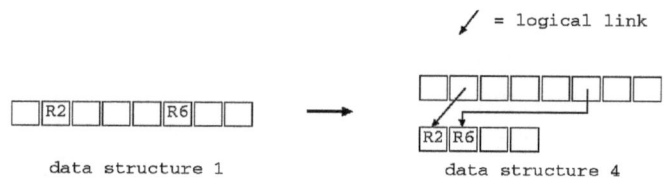

Fig. 4.7 Adding an array of pointers on top of an array of records. The assumption is made that maximum four records are stored in the data structure (at any moment in time)

4.4.3 Exploiting Indirection

In Fig. 4.7, a data structure consisting of one array (i.e., data structure 1) is transformed into a data structure consisting of two arrays (i.e., data structure 4).

Data structure 1 is once again the initial sparse array of records. Data structure 4 is the transformed data structure: an indirection layer (i.e., the top array) is introduced.

The top array, of data structure 4, contains the index values (or logical pointers) into the bottom array of data structure 4. Every array element of the top array corresponds to a unique record R. On the other hand, in the bottom array, every array element can potentially contain any record R.

Data structure 4 has a larger memory footprint than data structure 1. Analyzing Fig. 4.7 in terms of data accesses, we conclude that looking up a specific record R is more expensive for data structure 4 than it is for data structure 1 due to the indirection. The same holds for traversal: eight accesses are needed for data structure 1 while ten are needed for data structure 4 (i.e., eight accesses for the top array and two accesses for the bottom array).

To summarize, transforming data structure 1 without indirection into data structure 4 with indirection has the disadvantage that memory footprint and the data accesses to find and traverse increase. We will however frequently apply this transformation (together with other transformations) and obtain an overall decrease in access count. However, the traversal operation can be simplified to only traversing the records (R) and not the array elements of the top array since it is not necessary to know the key values in that particular context. This considerably reduces the number of data accesses during traversal.

4.4.4 Marking

In Fig. 4.8, a data structure consisting of one array (i.e., data structure 1) is transformed into a data structure consisting of two arrays (i.e., data structure 5).

Data structure 1 is the initial sparse array of records. Data structure 5 is the transformed data structure: an array of elements containing 0 or 1 is introduced. This top array is called a bit vector. An implicit correlation is present between the top array and the bottom array. For instance, since the value stored in the second array

Fig. 4.8 Adding an array of markings on top of an array of records

element of the top array contains a 1, the second array element of the bottom array (i.e., R2) contains a record.

In this specific example, the bit vector is only eight bits or one byte long. This implies that traversing the bit vector only consumes one data access as opposed to eight data accesses.

The total memory footprint of data structure 5 is only slightly larger than that of data structure 1.

Analyzing data structure 5 in terms of data accesses, we observe that, to insert a record, a total of three data accesses is needed: (a) one data access to store the record in the bottom array, (b) one data access to retrieve all the bits in the bit vector, and (c) one data access to store the updated bit vector. Note that for step (a) the entire byte of the bit vector needs to be retrieved. If the bit vector is more than one byte long, this implies that sometimes more than one data access is needed (i.e., the complexity is linear to the number of records stored in the bottom array).

Looking up a specific record in data structure 5 is achieved in one data access. Only the bottom array needs to be consulted.

Traversing data structure 5 is cheap. One data access is needed to retrieve the bit vector and consequently, two data accesses are needed to retrieve the two records R2 and R6 from the bottom array.

To summarize, transforming data structure 1 without marking into data structure 5 with marking has the disadvantage that memory footprint increases. However, the data accesses needed to traverse all records decreases. The number of data accesses needed to find a specific record remains unchanged.

4.4.5 Key Splitting

Our fifth data structure transformation is key splitting [52]. Since this is a more involved transformation, we defer the explanation to Sect. 4.5.7 in which we directly apply key splitting on an illustrative realistic example.

4.4.6 Partitioning

In Fig. 4.9, an array (i.e., data structure 1) is partitioned into two arrays (i.e. data structures 6a and 6b). This transformation allows a set of different sized records to be decomposed into two (or more) sets of similar sized records. As a result, after applying

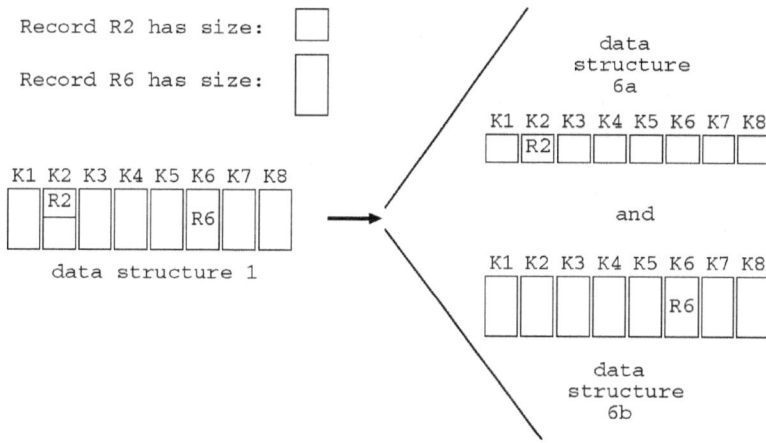

Fig. 4.9 Partitioning a data structure (e.g., an array) into two data structures (e.g., two arrays)

the transformations presented previously, on each set of similar sized records individually, we increase the search space of low cost data structure implementations considerably.

4.5 Example: DDT Transformations in Tetris Game

In this section, we illustrate the application of the previously presented DDT transformations to the Pixels Buffer of a Tetris game and give insights on the effects of each transformation.

4.5.1 Tetris Game

Tetris is a popular computer game in which the user controls falling objects of varying shapes and sizes. During game play, multiple objects are piled at the bottom of the screen and typically this pile grows until it reaches the top of the screen. When the latter occurs, the game ends. Multiple variants of the game exist, but we concentrate in particular on a version of the game in which multiple objects can fall down simultaneously while the user is interactively controlling the movement of these objects.

Each object has a specific shape which is decomposed into rectangles. A rectangle has one color and can be associated with a specific position on the screen by storing its x and y coordinates of the lower left and upper right corners. In this version of Tetris, for illustration purposes and without any loss of generality, we consider that the maximum number of rectangles is 100. We exploit this knowledge when transforming the Pixels Buffer.

4.5.2 The Pixels Buffer

The Pixels Buffer contains a data structure which manages the rectangles of the Tetris game. This includes the movement of the rectangles and the rendering of the rectangles from RGB format to video output (i.e., YUV chrominance-luminance) format [123].

We present a typical implementation in Sect. 4.5.3 and apply the transformations of Sect. 4.4 to obtain more economical implementations in terms of data accesses and/or memory footprint in Sects. 4.5.4 and 4.5.5. In Sect. 4.5.6 we compare the different implementations. In Sect. 4.5.7 we introduce and directly apply key splitting to the Pixels Buffer.

4.5.3 The Initial Pixels Buffer: A Sparse Array

A typical implementation, Implem1, of the Pixels Buffer's data structure is a sparse array of records. This data structure is presented in Fig. 4.10 in which, only three out of the maximum 100 rectangles are shown (to simplify the figure).

The position of the lower left corner of a rectangle corresponds to the index value of an array element. The record stored in the array element contains the position of the upper right corner and the RGB color of the rectangle.

The most important access operations are insertion of a rectangle, removal of a rectangle, and traversal through all rectangles (in order to render them onto the screen). Note that this knowledge can easily be obtained by profiling the Tetris game or by asking the programmer of the game.

In Table 4.1 we present the memory footprint of the array and the number of data accesses that are needed for the access operations under the (realistic) assumption that maximally 100 rectangles are present during game play. Since a Pos value and an RGB value both consume three bytes, we obtain a total memory footprint of $640 \times 480 \times (3 + 3)$ bytes or $1,800$ KB.

One data access is needed to insert a rectangle into the array. Since 100 rectangles need to be inserted, a total of 100 data accesses are needed. To remove a rectangle from the array, one data access is needed. This too amounts to 100 data accesses for the removal of 100 rectangles. To traverse through the array (i.e., to consult all 100 rectangles) a total of $640 \times 480 = 307,200$ accesses are needed.

4.5.4 Implem2: Explicit Keys

A second implementation, Implem2, of the Pixels Buffer is obtained by making the keys (or the indices) of the array of Fig. 4.10 explicit. This is in correspondence to Sect. 4.4.2 and results in the data structure of Fig. 4.11. The two arrays in this data

```
Pos(5,2) = lower left position of a tetris rectangle

Pos(7,1),RGB(blue) = upper right position and
                     color of a tetris rectangle
```

```
          ...              ...                    ↑
Pos(4,2)  [                          ]            |
                                                  |
Pos(5,2)  [ Pos(7,1),RGB(blue)       ]            |
                                                  |
Pos(6,2)  [                          ]            |
                                                  |
Pos(7,2)  [                          ]            |
                                                  |
Pos(8,2)  [                          ]            |
                                                    307200 entries
          ...              ...                    |
Pos(1,4)  [                          ]            |
                                                  |
Pos(2,4)  [ Pos(3,3),RGB(red)        ]            |
                                                  |
Pos(3,4)  [                          ]            |
                                                  |
          ...              ...                    |
Pos(5,6)  [ Pos(7,5),RGB(blue)       ]            |
                                                  |
Pos(6,6)  [                          ]            ↓
```

Fig. 4.10 Implem1: A sparse array of records. A record represents the position (Pos) of the *upper right corner* and the *RGB* color of a *rectangle* in the Tetris game. Each array element is indexed by a position value which represents the *lower left corner* of a rectangle. The total number of array elements is equal to the resolution of the screen, i.e., $640 \times 480 = 307,200$. Only 100 of these contain a Pos-RGB value since there are only 100 rectangles in our version of the Tetris game. Note that for simplicity only three of the 100 rectangles are shown

Table 4.1 Different implementations of the Pixels Buffer's data structure

	Memory foottprint (KB)	Insert	Remove	Traverse	Power (mW)
Implem1	1,800	100	100	307,200	368.9
Implem2	0.88	400	3,100	201	4.4
Implem3	300.9	500	900	201	1.9
Implem4	11.1	1,200	750	201	2.6

Memory footprint (KB), the number of data accesses to insert 100 Tetris rectangles, remove 100 Tetris rectangles and traverse once through all 100 Tetris rectangles, and the estimated power consumption (mW) are presented for one output frame

structure are implicitly correlated. This means that when for instance Pos(2,4) is retrieved from the first array, the associated Pos(3,3)-RGB(red) pair can be retrieved in the second array in a second data access.

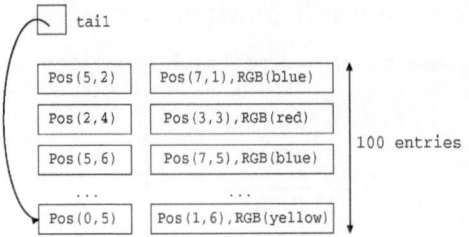

Fig. 4.11 `Implem2`: A more compact storage representation: the keys are explicit. Only four of the 100 rectangles are shown for simplicity

The memory footprint of this data structure is only $100 \times (3 + 6) + 1$ bytes or 0.88 KB. In this calculations, we realistically assume that both a Pos value and an RGB value consume three bytes. Also, the tail pointer consumes one byte.

Inserting a rectangle only takes four data accesses: (a) one data access to retrieve the tail pointer's value, (b) two data accesses to store the Pos and Pos-RGB values respectively, and (c) one data access to store the incremented tail pointer. Thus, for the insertion of 100 rectangles, a total of 400 data accesses are needed.

Removing a record is a more involved operation. It includes (a) finding the to-be-deleted Pos and Pos-RGB entries (which takes a non constant amount of accesses). Assuming that in the average case 50 rectangles are stored in the data structure, a total of $1 + \frac{50}{2} = 26$ data accesses are needed. One access is needed for the tail pointer and 25 accesses are needed in the average case to find a random rectangle in the data structure. In addition to this, the following needs to be done. (b) Copying the last Pos and Pos-RGB values into the previously found entries in four data accesses, and (c) storing the decremented tail pointer in one data access. In total this amounts to 31 data accesses. For 100 rectangles, a total of 3,100 accesses are needed (to implement the removal of those 100 rectangles). This is about a factor three more accesses than in `Implem1`.

However, traversing the data structure (when all 100 rectangles are stored) is achieved by (a) retrieving the tail pointer's value in one data access and (b) accessing each Pos and Pos-RGB entry in 200 data accesses. This results in a total of 201 data accesses for traversal. This is about 1,500 times less accesses than in `Implem1`.

4.5.5 Implem3: Explicit Keys and Indirection

A third implementation of the Pixels Buffer is obtained by adding a layer of indirection to `Implem2`. This is in correspondence to Sect. 4.4.3. The data structure is shown in Fig. 4.12. In this implementation, the key values (i.e., the positions of the lower left corner of the rectangles) are stored both implicitly (in array 1) and explicitly (in array 2). This form of redundancy results in (a) a large memory footprint but (b) a small number of data accesses for the removal of a rectangle as is shown in Table 4.1.

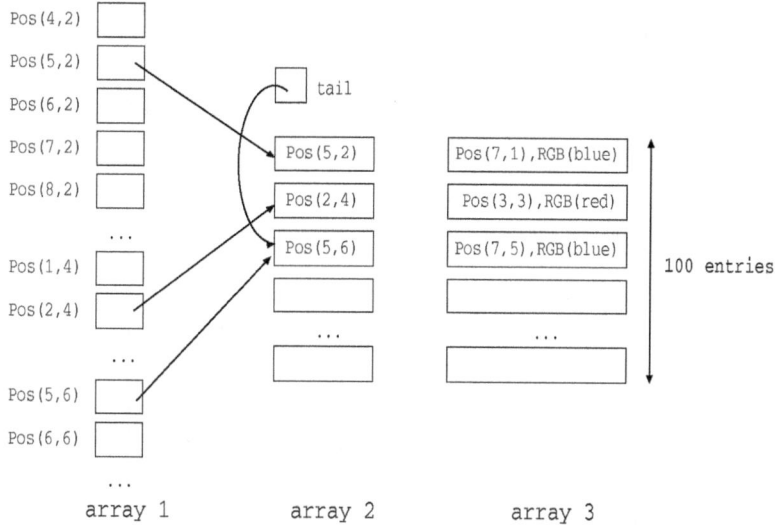

Fig. 4.12 Implem3: A storage representation where the key values are stored both implicitly (array 1) and explicitly (array 2). The drawing corresponds to the scenario where only three of the 100 rectangles have been inserted into the data structure. This is done for simplicity. Other 97 rectangles still need to be inserted

To insert a record, an additional data access (for array 1) is needed in comparison to Implem2. This results in a total of five data accesses for one rectangle and thus 500 data accesses for 100 rectangles.

Removing a record consists of the following steps: (a) the to-be-deleted Pos and Pos-RGB entries are found via array 1 in one data access, (b) the indirection (i.e., the pointer from array 1 to array 2) is deleted in one data access, (c) the tail pointer is consulted in one data access, (d) the last Pos and Pos-RGB values are copied into the previously found entries (of arrays 2 and 3) in four data accesses, (e) the indirection from array 1 to these moved Pos and Pos-RGB values is updated in one data access, and (f) the decremented tail pointer's value is stored. This takes nine data accesses in total. For 100 rectangles, a total of 900 data accesses are needed. This is a factor three less accesses in comparison to Implem2.

Traversing through the data structure is similar to Implem2: only arrays 2 and 3 are used. Array 1 in Fig. 4.12 is not used during traversal. Thus, a total of 201 data accesses are needed.

4.5.6 Comparing the Different Implementations

Comparing Implem1, 2, and 3, we observe that the least power consuming implementation is Implem3. On the other hand, Implem2 consumes the least amount of memory footprint. Note that in Implem2 and Implem3, a total of 201 data accesses

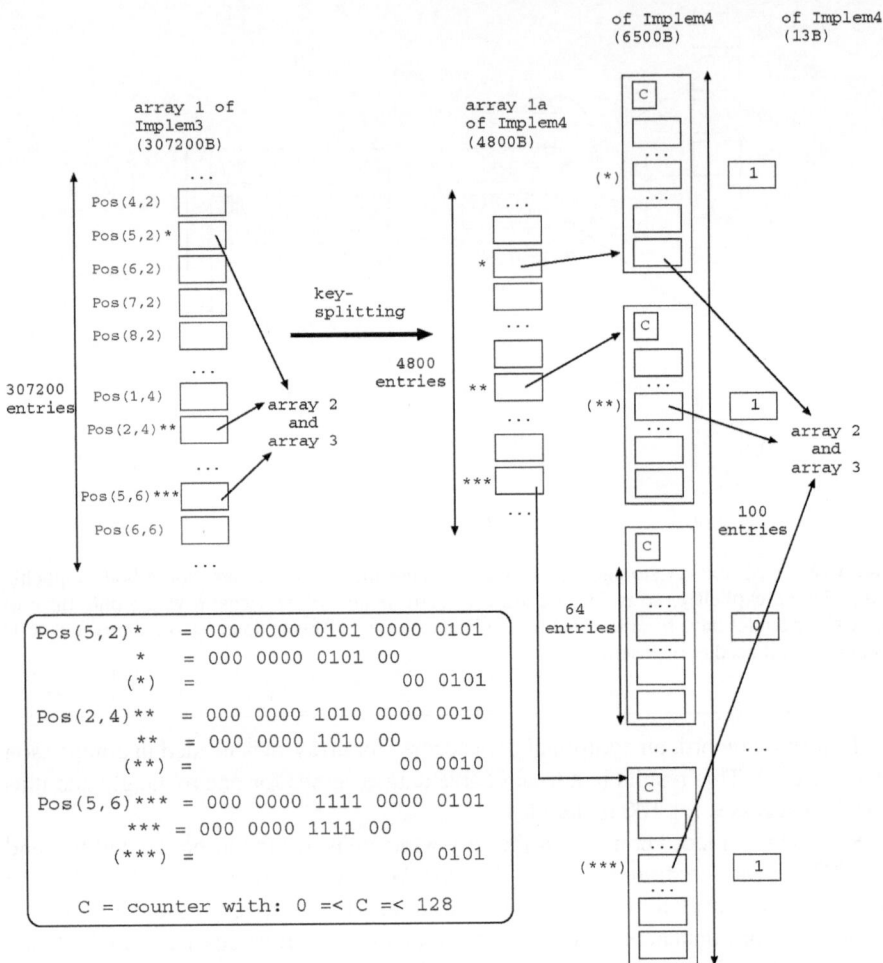

Fig. 4.13 Key splitting of array 1 of `Implem3` into three arrays that belong to `Implem4`. The arrows leaving the small array elements in array 1b point to specific entries of array 2 (cf., Fig. 4.12)

are needed for traversal. This is because both data structure implementations use the same arrays (i.e., arrays 2 and 3 and the tail pointer in Fig. 4.12) for traversal.

In Sect. 4.5.7 we present another transformation and apply it to `Implem3`. This will result in `Implem4` (see last row in Table 4.1) which has a decreased memory footprint and only a small increase in overall access count (and hence power).

4.5.7 *Implem4: Key Splitting*

In this section we apply key splitting to array 1 of `Implem3` (see Fig. 4.13). Key splitting implies that the key bits, that make up a key value, are split into two (or more) groups. For instance, the Pos(5,2) value is a key value present in array 1 of Fig. 4.13. In binary form this corresponds to 000 0000 0101 0000 0101. A possible key splitting is to split these 19 key bits into one group of 13 bits and another group of six bits, i.e., 000 0000 0101 00 and 00 0101 respectively. The newly obtained 13 key bits represent the key value of an entry in array 1a in Fig. 4.13. Similarly, the newly obtained six bits represent the key value of an entry in array 1b. The reason why we have chosen to split the 19 key bits into respectively 13 and six key bits is given below.

Let y represent the number of key bits of an array element of array 1b; e.g., $y = 6$ in the above example. Let N represent the total number of array elements of array 1a. The resolution of the video screen is 307,200 pixels. Based on these definitions, we know that $N \times 2^y = 307,200$. Choosing a specific value for y (e.g., $y = 6$), results in a value for N (e.g., $N = 4,800$). The total number of key bits for an array element of array 1 is 19. Therefore, $19 - y$ represents the number of key bits for an array element of array 1a. In our particular example, $19 - y = 13$ key bits.

The objective of key splitting is to reduce the total memory footprint. In other words, the total memory footprint of arrays 1a and 1b in Fig. 4.13 is much less than the memory footprint of array 1. We calculate the memory footprint of arrays 1a and 1b as follows: MemFootpr $= N \times 1$ byte $+ 100 \times (2^y \times 1$ byte $+ 1$ byte). The factor 100 is present in the equation because we know that a maximum of 100 Tetris rectangles are present in the Tetris game. Each large array element of array 1b has a counter C whose size is one byte. It denotes the number of stored entries. Since we want to minimize the memory footprint, we apply the previous two equations for different values of y and select the value which corresponds to the minimum memory footprint. In this example, the best value of y is $y = 6$ and consequently $N = 4,800$. The memory footprint of arrays 1a and 1b together is 11.1 KB. Compared to the memory footprint of array 1, which is 300 KB, this is a significant reduction. In Fig. 4.13, these optimal values are used.

In addition to arrays 1a and 1b, we add an additional array 1c as well. Each array element of this array marks (cf., Sect. 4.4.4) a large array element of array 1b. Recall that $2^7 = 128$ small array elements are present in each large array element of array 1b; i.e., array 1b is a two dimensional array in programming jargon. If no small array elements are used in a specific large array element of array 1b, then the corresponding array element of array 1c contains a 0, else it contains a 1. Since there are 100 large array elements of array 1b, the same amount of array elements of array 1c are present as well. This implies that the memory footprint of array 1c equals $\frac{100}{8} = 13$ bytes.

Even though the total memory footprint of arrays 1a, 1b, and 1c is significantly smaller than the memory footprint of array 1, the data accesses needed, to find, insert, or retrieve have increased. For instance, finding a specific record implies that array 1a is consulted first by specifying the thirteen most significant key bits, then array

2a is consulted by specifying the six least significant key bits, and finally the record is retrieved in arrays 2 and 3. Even more access overhead is present for insertion and retrieval but we omit further explanation for brevity. The numerical results are shown in Table 4.1.

We can observe in Table 4.1 that `Implem4` lies in between two extremes, namely `Implem2` and `Implem3`. `Implem4` has a relatively small access count and a relatively small memory footprint in comparison to the other implementations. This makes `Implem4` a very interesting implementation point for certain situations. It also forms another Pareto point in the overall trade off space. Also recall that many key splitting variants of `Implem4` can exist (including implementations in which key splitting is applied more than once) which are additional Pareto points in the Pareto space. We have only presented one possibility in the Tetris example to illustrate this transformation.

4.6 Exploration and Optimization Methodology of DDTs Implementations

From the point of view of the modular programming paradigm and the use of abstract data types [5, 32], which is used in all modern programming languages, the functionality of any program should be decoupled and defined independently from the specific implementation of the data structures (dynamic or static) used to store the application data, as long as all the methods and access functions required by the implemented algorithm are available [82]. According to this software programming paradigm, the proposed methodology relies on the assumption that the DDTs can be identified in the original source code of the application under exploration, and they can be replaced without collateral effects that require the modifications of the control flow in the application. In other words, the input source code of the application where the presented methodology is applied must not include the DDT implementations, and they must have been defined in a separate module or standard library of DDT implementations in C++, such as the Standard Template Library [145, 157]. Moreover, all the algorithm using the DDTs implementations must interact with them only through the standard and common set of access methods [6] (see Sect. 4.6.2 for more details).

It must be noted that the aforementioned restriction does not impose any real limitation with respect to the applicability of the methodology, as all the studied applications and most of the applications nowadays, specially designed using the object-oriented paradigm, are implemented following this assumption. Moreover, as the object-oriented programming paradigm is being adopted further as the common design standard to exploit the System-on-Chip paradigm at the software level, this previous assumption will need to be enforced further, due to the necessity to define and use reusable architectures, based on standard interfaces.

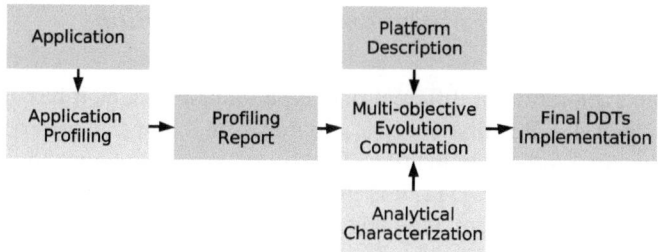

Fig. 4.14 Overall DDTs optimization flow

The proposed exploration and optimization methodology uses three different phases to perform the automatic exploration of DDT implementations using evolutionary computation. Figure 4.14 shows an overview of the different phases (in light gray) and the inputs (in dark gray) required to perform the overall DDTs design exploration and optimization. In the first phase, there is an initial profiling of the iterator-based access methods to the different DDTs used in the application (Sect. 4.6.1). To this end, the presented profiling framework presented in Chap. 3 must be plugged into the source code under exploration. This means that the declaration of the DDTs in the application must be altered with the declarations used by a profiling library. This initial step is the only one that breaks the automated chain of exploration, but it is manageable in most modern applications, as only one line of source code must be modified for each variable to be explored using the proposed methodology.

Next, using this detailed report of the accesses to the DDTs done by the application, by using our own library of basic and multi-level DDTs (Sect. 4.6.2.2), defined according to the previously presented transformations targeting multimedia and network applications for nomadic embedded applications (Sect. 4.4) and the characteristics of the final platform, an exploration of the design space of DDTs implementation is performed using multi-objective evolution computation (Sect. 4.6.3). As a result of this second phase of our exploration methodology, Pareto charts are drawn, according to the design constraints. Every point of the Pareto chart corresponds to a different combination of DDT implementations for all the variables explored in the application. The log file is used as an input to obtain Pareto charts between memory accesses, memory footprint, performance and energy consumption. Thus, the designer can selects in the developed tools different trade-offs between these factors by selecting different points from the Pareto optimal curve.

Once the designer has selected a Pareto optimal implementation option, the code of the set of multi-layer DDT implementations for the explored variables of an application can be automatically generated and returned to the user to be used directly in the target nomadic embedded application.

4.6.1 Profiling Library of Dynamic Data Allocation

To enable the exploration of different data type refinements, as it has already been explained in Sect. 4.2, it is first necessary to understand how the different DDTs are being used in each studied application. To this end, since the target applications are dynamic, hence the use of DDTs, it is therefore necessary to profile them to get an accurate view of the different demands of the data types at run-time. As we want to explore the different potential implementations for the DDTs, it is necessary that this profiling happens not at the memory level, but at the interface level. Thus, we have developed our own profiling library [47] with several higher level profiling packets, which is presented in detail in Chap. 5. This profiling library relies on the standard interface defined by STL [145], where we have re-implemented a sequence type, vector, that logs all the different semantical operations. As indicated in the previous section, a practical advantage to sticking to a commonly used interface, such as STL, is that limited changes are required in the algorithms to profile an application, and they can be performed automatically by searching for the declaration of data types in the code sources of the application with a standard parser of a C++ compiler, and then changing the type of the found data types from the STL vector-type to the profiling vector-type, without requiring a modification of the remainder of the application where the data type is actually being accessed.

A careful analysis of the sequence interface indicates that not only operations of the DDTs or containers, but also the iterator operations used to access the stored elements [8, 134] must be logged. To enable us to couple the logging of memory accesses to specific containers, it is necessary to know at each point in time, from the profiling information, which container uses which memory segments. Therefore, the constructor, destructor, copy constructor and swap operation are logged as separate packets. Other such operations are the accessing of an element, the addition of an element, the removal of an element and the clearing of the container. Since it is possible to obtain references to an element in a container, no distinguishing exists between reads and writes. Additionally, the iterator methods to access an element or the updating of an iterator, either incrementally or using random offsets, are distinguished as well. The reasoning to differentiate these two types of operations is that this gives different trade-offs for the implementation of a DDT, which need to be explored in our optimization process using evolutionary computation (see Sect. 4.6.3). For instance, for an array-like data type, randomly moving the iterator through the contents of the array has $O(1)$ access and large $O(n)$ memory footprint requirements, while for a compact list-like implementation of a sequence, this has $O(n)$ access requirements. A complete description of the implementation details and design approach of the profiling library used in our methodology is presented in Chap. 5.

4.6.2 Library of DDTs Implementations for Multimedia and Communication Applications

When analyzing DDTs, we divide them into two basic types: arrays and graphs. An array contains values (i.e., scalar, complex or even abstract ones when storing pointers to other data). The most important property of an array is that the entire structure is allocated on one shoot, but the memory can be allocated at compile time (static memory) or at run time (dynamic memory). On the other hand, graphs consist of multiple nodes that can store a scalar or complex value each. Furthermore, it contains at least one pointer to another similar memory location (that can be NULL). The combination of all connected nodes forms the complete graph. Contrary to arrays, nodes in a graph are allocated and freed incrementally: when more data needs to be stored, additional nodes are allocated. In most cases, two special cases of graphs are used: trees and lists. In a tree, each node has one *parent* node and at least one *child* node. Exceptions are the *root* node that has no parent and the *leaf* nodes that have no children.

Complex layered data types are combinations of the two aforementioned basic types. In a typical example, the overhead memory in linked lists (i.e., the pointers to the next and previous nodes) are amortized by allocating memory for several elements at once. In this case, the first layer is a linked list and the second one is an array. As a result of the possible combinations of basic types in complex layered structures, the search space of these complex data types grows exponentially and a systematic exploration and construction method becomes a must.

In the proposed methodology it is particularly necessary to be able to construct the most suitable DDT implementation for variable included in the multimedia and network application under study. This implementation will be a multi-layered structure, which is the result of applying one or several of the transformations (Sect. 4.4) to the initial (and more basic) dynamically-allocated data structure, namely, a dynamic vector, which is used in the application profiling phase. Thus, the main challenge in this context is the development of a flexible method to apply each of the proposed transformations, minimizing the effort of creating new DDT implementation. Furthermore, this effort should be independent from the particular application under study and, hence, the DDT implementations must be reusable between application.

In order to fulfill all the previous requirements, we have developed a library of object-oriented DDT implementations in C++, which using a C++ modeling approach based on template C++ classes [166] and mixins [150]. In the following sections, this elementary concept is described and the complete library of DDT implementations is described, as well as the common access methods to all the DDT implementations.

```
template <class SupClass> class Mixin:
    public SupClass{ // mixin definitions };
template <class SupClass> class TemplateClass{
    SupClass* data;
    // template class definitions };
```

Fig. 4.15 Parametrized Inheritance used with mixins in C++

4.6.2.1 Modular Composition of Multi-layer DDT

The use of template C++ classes [166] and mixins to construct DDT implementations allows easy and flexible modeling, and refinement of layered DDTs. In the remainder of the text, we use the definition of mixins as used in [150]: a method of specifying extensions of a class without defining up-front which class exactly it extends. In C++, a subclass is specified and the parameterized parent class is determined during the instantiation of the code later on.

Figure 4.15 declares a subclass of SupClass with SupClass itself being a template argument. Since we are using the inclusion method [166], also the subclass is defined. The Mixin is written for one or more types that are not yet specified. A small variant can be seen in the second class definition (i.e., TemplateClass), where the template argument is not used as a parent class, but instead as internal private data members. We use the second approach of template classes to model the memory behavior of DDT implementations, while the mixin approach is used to add, modify and refine functionality of complex DDTs.

In our library, we have implemented the typical primitive data structures used in multimedia and network applications [138, 175] using our mixins-based layered fashion. An example of these simple DDTs are shown in the middle part of Fig. 4.16. Two different sorts of simple lists are implemented, single linked (i.e., SLL) where each node is pointing to the next one only and double linked (i.e., DLL) where every node points to the next and previous node. Also, a binary tree (i.e., BTTree or BTT) and array (Array) are implemented. All of them can include default interface methods like GetElement, AddElement, DelElement, etc. These simple DDTs are the only DDTs that need to be specified/written to compose more complex ones in dynamic multimedia and network applications. Then, to use the template code the developer just needs to specialize them. For example, Fig. 4.16 defines a number of base classes: an array of 256 float elements, a double linked list of ints and a binary tree of doubles. In these cases, the last parameter is a combination of two mixins (first line in Fig. 4.16): First, mheap is a thin wrapper for the memory operations malloc() and free(). Second, TypeClass is used as intermediate layer to define the basic data types of the layers (i.e., float and int).

Next, these basic DDT layers can be combined in complex multi-layered DDTs as shown in Fig. 4.17. In the first case of Fig. 4.17, a double linked list of arrays of 128 integers is declared (i.e., DLLAR). Also, a binary tree of a single linked list is defined (i.e., BTSLL), followed by a multi-layered array structure (i.e., ARARAR) with an

```
template<type T> class Mem:public TypeClass<T,mheap>{};
template<int Size,type T,class SupClass> class Array{..};
template<type T,class SupClass> class SLL {...};
template<type T,class SupClass> class DLL {...};
template<type T,class SupClass> class BTTree{...};
class F_Array: public Array<256,float,Mem<float> >{};
class I_DLList: public DLLList<int, Mem<int> >{};
class D_BTTree: public BTTree<double, Mem<double> >{};
```

Fig. 4.16 Basic Abstract Data Types definitions and instantiations

```
class DLLAR: public \
    DLL<int,ArrayClass<128,int,Mem<int> > >{};
class BTSLL: public \
    BTTree<double,SLL<double,Mem<double> > >{};
class ARARAR: public \
    ArrayClass<2,Point3D,ArrayClass<4, Point3D,\
        ArrayClass<2,Point3D,Mem<Point3D> > > >{};
```

Fig. 4.17 Examples of multi-layered DDTs

array of 2, pointing to an array of 4, pointing to arrays of 2 `Point3D` elements. These can be typical custom DDTs of a real multimedia application [138].

4.6.2.2 Elementary Set of DDT Implementations for Multimedia and Communication Applications

This section describes the different options of the library of basic and multi-layered DDT implementations for multimedia and network applications. More specifically, we describe in the next paragraphs their main functionality and the analytical characterization of their features (in terms of memory footprint and memory accesses in case of sequential and random accesses).

As it has already been mentioned, a DDT is a software abstraction by means of which we can manipulate and access data. The implementation of any multi-layered DDT has two main components. First, it has storage aspects that determine how data memory is allocated and freed at run-time and how this memory is tracked. Second, it includes an access component, which can refer to two different basic access patterns: sequential or iterator-based and random access. In our case we have classified the DDT implementations that can cover the previously defined data structure transformations (Sect. 4.4) in basic DDT and multi-layer implementations, as proposed in [47]. The basic DDTs are the following ones:

- `Array` (AR): is a set of sequentially indexed elements of size s_T. Each element of the array is a record of the application.
- `Single Linked List` (SLL) and `Double Linked List` (DLL): SLL (Fig. 4.18) is a single linked list of vectors of type s_T. Each element of the list is

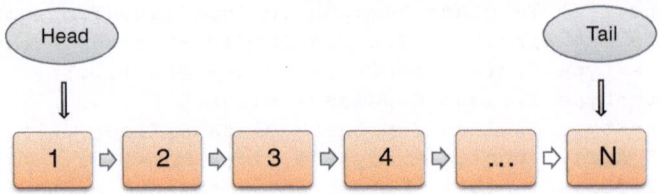

Fig. 4.18 A single linked list—SLL

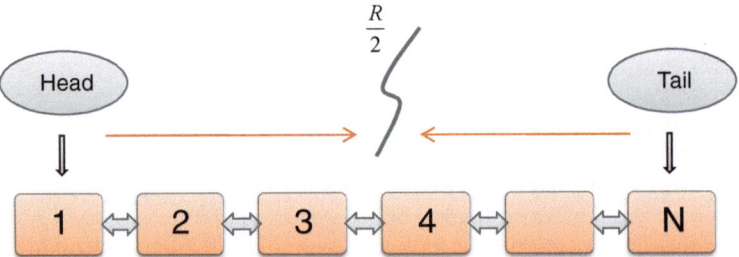

Fig. 4.19 A double linked list—DLL

connected with the next element through a pointer, while DLL (Fig. 4.19) is a double linked list of vectors of type Te. Each element of the list is connected with the next and the previous element with two pointers (of size s_w), one pointing to the previous element and one to the next.

From a theoretical point of view, a list is a homogeneous collection of elements, with a linear relationship between elements of type Te. Linear means that, at the logical level each element in the list except the first on has a unique predecessor, and each element except the last one has a unique successor.

Lists may be implemented in a variety of ways, each one satisfying the condition of linearity. Nevertheless, in the implementation level, a relationship also exists between elements, but the physical implementation may not be the same as the logical one. For example, in a double linked list, each element is connected to both the next and the previous one. Such variations in the implementation level of the Abstract Data Type List provide each data structure with unique characteristics as far as the number of accesses of each operation is concerned.

Lists can be either Sorted or Unsorted. In the first case their elements may be placed into the list in no particular order, while in the second case they can be sorted in a variety of ways. For instance, a list of numbers can be sorted by value, a list of strings can be sorted alphabetically, and a list of grades can be sorted numerically. When the elements in a sorted list are of composite types, their logical order is determined by one of the members of the structure, called the key member. This is an analog concept to the key field to access an element in the DDT transformations presented in Sect. 4.4 (Fig. 4.6).

Table 4.2 Analytical characterization of DDT implementations considered in the exploration

DDT implementations	Sequential (NA_s) access count	Random (NA_r) access count	Average size (S_{av})
SLL(AR)	$3 \times N_e + N_a$	$\frac{N_e}{2 \times N_a} + 3$	$5s_w + \frac{N_e}{N_a}$
SLL(ARO)	$3 \times N_e + N_a$	$\frac{N_e}{2 \times N_a} + \frac{3 \times N_a}{N_e} + 1$	$7s_w + \frac{N_e}{N_a} \times (3s_w + s_T)$
DLL(AR)	$3 \times N_e + N_a$	$\frac{N_e}{4 \times N_a} + 3$	$6s_w + \frac{N_e}{N_a} \times (4s_w + s_T)$
DLL(ARO)	$3 \times N_e + N_a$	$\frac{N_e}{4 \times N_a} + \frac{N_a}{5 \times N_e} + \frac{5}{4}$	$8s_w + \frac{N_e}{N_a} \times (4s_w + s_T)$
SLL	$3 \times N_e$	$\frac{N_e}{2} + 1$	$4s_w + N_e(2s_w + s_T)$
SLL(O)	$3 \times N_e$	$\frac{N_e}{2} + \frac{1}{N_e}$	$6s_w + N_e(2s_w + s_T)$
DLL	$3 \times N_e$	$4N_e + 1$	$5s_w + N_e(3s_w + s_T)$
DLL(O)	$3 \times N_e$	$\frac{N_e}{4} + \frac{2}{N_e} + \frac{1}{4}$	$7s_w + N_e(3s_w + s_T)$
AR	N_a	1	$N_a \times s_T$
AR(P)	$2 \times N_a$	2	$N_a(s_T + s_w)$

In addition, we have included in the library of DDT implementation the fundamental variations of basic DDTs, resulting from the application of the transformations presented in Sect. 4.4, which are relevant for embedded multimedia and network applications [10, 21, 174], namely:

- Pointer (P): in the pointer variation of each basic DDT, the record of the application is stored outside the DDT and is accessed via a pointer. This leads to a smaller DDT size, but also to an extra memory access to reach the actual data. All DDTs used in our exploration comply to this variation except the simple array.
- Roving Pointer (O): The roving pointer is an auxiliary pointer (of size T_{ref}) useful to access a particular element of a list with less accesses in case of iterator-based access patterns. For instance, for an array if you access element $n + 1$ immediately after element n, your average access count is $1 + 1$ instead of $n/2 + 1$.

Then, these simple DDTs can be combined in multi-layer structures that offer different trade-offs between memory use, performance and energy consumption. These trade-offs are shown in the analytical characterization of the multi-layer DDTs used in our exploration, presented in Table 4.2. This analysis is based on a number of parameters, which can be extracted from the initial profiling phase of the presented method, namely; N_e is the number of elements in the DDT, N_a is the number of elements stored in the DDT, s_w is the width of a word on the architecture and s_T the size of one element of type T, and grouping-level of the basic elements (FA). These equations are used in the evaluation phase of our exploration process of DDT implementations using evolutionary computation, which is explained next (Sect. 4.6.3), since dynamic embedded multimedia and network applications are made of multiple dynamic variables and the use of a concrete DDT implementation for a particular variable can affect the overall cost of the rest of the variables in the multi-objective optimization process.

Table 4.3 Common set of data access public/private operations in all the DDT implementations of the proposed library, provided by the SETgeneric public class

Public operations	Method provided by SETgeneric
constructor	SETgeneric()
add element	virtual T * addelement(T * element)
destructor	SETgeneric()
nr of elements	virtual unsigned nadds()
return element	virtual T * getelement(unsigned position)
delete element	virtual T * getelement(unsigned position)
last element	virtual bool popelement(T * outelement)
first element	virtual bool firstElem(T * outelement)
traverse-iterator	virtual bool collect(T * data)
Private Operations	Method provided by SETgeneric
pointer element	virtual T * getelementpointer(unsigned position)
delete all data	virtual bool deldata()

4.6.2.3 Set of Access Methods to DDT Implementations

Although the implementations of the multi-layer DDTs included in our library for multimedia and network applications are very different, in order to facilitate the addition of the library to any application, all the DDTs provide a common data access interface. This interface is provided by an additional abstract data class, i.e., SETgeneric. This is the only visible class for application software engineers and covers all the necessary interface to access the stored data in multimedia and network applications. The equivalence between the public access methods provided by SETgeneric is shown in Table 4.3.

As it is shown in Table 4.3, apart from the set of public methods or operations provided for the application software engineer, such as, add/delete an element to manage the DDT, or first/last to access particular stored elements, SETgeneric provides as well a set of private methods for the DDT implementation developer. Thus, it is possible to implement efficiently usual operations in multimedia and network applications. The two private operations are getelementpointer() y deldata(). On the one hand, getelementpointer() provides the means to efficiently modify an internal element of the DDT through a pointer to that particular elements. This operation is conceptually equivalent to call sequentially the operations: getelement(), delelement() and finally addelement(). However, thanks to the support provided by the C++ language to efficiently handle pointers (due to the semantics inherited from its ancestor C language) and due to the amount of times that this type of operation is performed in the latest multimedia and network applications, it is more efficient to provide this kind of implementation for this operation for the DDT developers. Hence, the access time and number of memory accesses can be largely reduced with respect to pure object-oriented operations. On the other hand, the deldata() operation is an enhancement of the delement() operation, which can be used to delete all the stored elements in a certain DDT just with a single function call, but without deleting the created object. Therefore, intermediate and frequent removals of intermediate data generated by multiple sub-algorithms of a certain application (see Sect. 4.3 for a real example) can be executed in an optimal way.

In the ideal case of application of the proposed methodology, the application software designer starts from the beginning of its development of a certain application directly using the set of operations and methods provided in Table 4.3. Thus, the profiling of the run-time behavior of the dynamically allocated variables of the application under development and the exploration of the optimal DDT implementation for each of them can operate in a completely automated way. Otherwise, the initial declaration of the variables of the target application that will be optimized with the proposed design methodology would need to be manually modified in the source code of the application.

Finally, as we have already verified in multiple commercial case studies [59, 71, 91, 116, 122, 124], the proposed set of data access methods (Table 4.3) is adequate for multimedia and network applications characteristic of nomadic embedded systems. However, if a different application domain is targeted, the set of operations would need to be revised and (if needed) adjusted, according to the representative access methods of the new types of target applications.

4.6.3 Multi-objective Exploration of DDT Implementations

The multi-objective exploration and optimization phase for the DDTs of multimedia and network applications of nomadic embedded systems relies on the previously explained analytical pre-characterization of the possible DDT implementations of our library, which are then used in a *Genetic Algorithm* (GA) of type *Vector Evaluated Genetic Algorithm* (VEGA) [143] to model the existing inter-dependencies of using different DDTs implementations. Then, this modeling of inter-dependencies can be seen as a constraint set that can be used to prune the design space. Moreover, this pruning method can select the best choice according to each designer's metrics. Hence, given an application to be optimized for a certain architecture of a nomadic embedded system, the proposed optimization framework can return the best multi-layer DDT implementation. This exploration is performed in a completely automated way, for each of the included dynamic variables in the target application. The optimization can take place either for a concrete designer-defined multi-objective optimization metric (i.e., memory footprint, memory accesses, energy consumption or linear combinations of them), or a number of overall solutions that respect the defined user constraints (Pareto front).

GAs [75] are stochastic optimization heuristics where the exploration of the solution space of a certain problem is carried out by imitating the population genetics stated in Darwin's theory of evolution. Selection, crossover and mutation operators, derived directly from natural evolution mechanisms, are applied to a population of solutions, thus favoring the birth and survival of the best solutions. GAs have been successfully applied to many NP-hard combinatorial optimization problems and work by encoding potential solutions (individuals) to a problem by bit strings (chromosomes), and by combining their codes and, hence, their properties. In order to apply GAs to a problem, a genetic representation of each individual has to be found. Furthermore, an initial population has to be created, as well as defining a cost function to measure the fitness of each solution. Then, we need to design the genetic operators that will allow us to produce a new population of DDT solutions from a previous one, by capturing the inter-dependencies of the different DDT implementations working concurrently. Next, by iteratively applying the genetic operators to the current population, the fitness of the best individuals in the population converges to targeted solutions, according to the metric/s to be optimized and the weight of each of these metrics. For a detailed overview of GAs the reader is referred to [118].

4.6.3.1 Genetic Representation

In order to apply a GA correctly we need to define a genetic representation of the design space of all possible DDT implementation alternatives. Moreover, to be able to apply the VEGA optimization process and cover all possible inter-dependencies of DDT implementations for different dynamic variables of an application, we must guarantee that all the chromosomes represent real and feasible solutions to the problem and ensure that the search space is covered in a continuous and optimal way. To this end, we define the implementation of the variables of a program by storing the following information on each chromosome:

- Data Structure (DS): this field represents one the 16 different possibilities using the previous DDTs analytically characterized (Table 4.2). Therefore, using a binary encoding we need 4 bits.
- Number of elements (Elements): this field represents the grouping factor of elements, up to eight elements (3 bits), which can create optimal access patterns in dynamic multimedia and network applications, as outlined by [52].
- Number of Levels of the Data Structure(Levels): this field can specify up to 4 levels of basic DDT implementations grouped together (2 bits).
- Basic Fields: Our methodology enables in its encoding the representation of up to 16 fields (4 bits) to cover a large exploration space in this field. However, according to our experience with real-life applications, we typically would not find optimal data structures with more than 7 or 8 basic fields, which can enable reducing the size of this field to improve further the exploration time. Nevertheless, even with this large size for this field, the observed exploration time for real-life applications is very limited, i.e., few hours (see Sect. 4.7).

Consequently, using this chromosome structure we need $(4 + 3 + 2 + 4) = 13$ bits to represent the solution proposed for each variable, and if an application has N variables, each chromosome has to be constituted by N*13 bits (genes). For example, the first application we have tested in our experimental results (i.e., Simblob,

0 to 1	2 to 5	6 to 8	9 to 12	13 to 14	15 to 18	19 to 21	22 to 25	Bit positions
Levels	Basic Fields	Elements	DS	Levels	Basic Fields	Elements	DS	Meaning
Variable 1				Variable 2				

Fig. 4.20 Example of a 26-bit chromosome

see Sect. 4.7) uses only two dynamic variables. A potential solution would be represented by a 26-bit chromosome (see Fig. 4.20). Our current implementation of the exploration framework is able to explore applications with up to 40 DDTs at the same time, which can cover all the real-life embedded applications we are aware of.

4.6.3.2 Fitness Function

After performing once the profiling of the real application for a realistic input set, the information required for the analytical characterization of the DDTs implementations considered is available (see Sect. 4.6.2.2 for more details), as well as the number of read (N_r) and write (N_r) accesses to each DDT during execution. Thus, for each DDT implementation available in a certain generation we can compute the performance (Perf) related to the number of accesses to memory), memory footprint (AvMem in Bytes) and energy values (AvMem in Bytes) using the equations of Table 4.2, according to the different types of access methods (sequential and random) used in each target embedded application. These values are evaluated for each individual DDT, which represents a solution, by means of a fitness function. The objective of our algorithm is to obtain DDT implementations which optimize the aforementioned metrics for each variable in the application (energy, memory use and performance). Therefore, the fitness process starts with the decodification of the individuals. Next, for each possible variable (and its valid DDTs in the current generation of the exploration) we compute the following equations:

$$Perf = (NA_r * (3 * (N_r + N_w - 2)/4))$$
$$+ (NA_s * ((N_r + N_w - 2)/4)) + (NA_{cd} * 2) \tag{4.1}$$
$$AvMem = S_{av} \tag{4.2}$$
$$Energy = ((Npa * Epa) + (N_{rw} * Erw)$$
$$+ (S_{av} * Eestatica))/1000 \tag{4.3}$$

where NA_r represents the number of accesses to the variable in the application, NA_s is the cycle time cost of an iterator-based access, NA_cd is the cycle time cost of creating/destructing the DDT implementation, and S_{av} is the average memory footprint of each variable, as calculated in Table 4.2. Regarding the energy calculations, in this book it is assumed to be used in-place sharing, as their lifetimes are short, and a basic memory hierarchy that consists of a main shared memory and a L1 data-cache. Other memory hierarchies could be modeled as well by modifying the previous equations. The result is obtained in nJ. In this case, N_{pa} is the number of misses in the L1 data-cache. E_{pa} is the energy consumed per access to main memory, N_{rw} is the number of reads/writes to the cache memory and E_{rw} is the energy consumption per access to the cache.

According to our empirical validation with several multimedia and network applications [47], it can be assumed in the energy calculations an average miss rate of the cache memory below 5 % of the overall memory accesses. However, this value is user-configurable in the proposed VEGA-based exploration process and even additional multi-level cache miss rate effects can be configured. In addition, it is possible to introduce some constraints and weights for the metrics to be optimized. For example, the maximum values of performance, memory use and energy can be fixed if the final embedded system requires it.

4.6.3.3 Multi-objective Algorithm

Multi-objective optimization could be defined in our case as the problem of finding a vector of decision variables which meets a set of constraints, and then this vector of decision variables is used to optimize a vector function whose elements represent the objective functions. These functions form a mathematical description of performance criteria which are usually in conflict with each other. Hence, the term optimize means finding such a solution which would give acceptable values to all the objective functions (energy, performance and memory in our problem) for the designer [131]. The notion of acceptable values is defined by the weight that the designer gives to each optimization metric, enabling linear combinations of the aforementioned metrics in our case and creating Pareto curves of solutions. In order to find these Pareto's curves for problems of great difficulty, several multi-objective evolutionary algorithms have been recently proposed [41]. Among them, one that has been demonstrated to be very efficient is the approximation proposed by Schaffer [143]. The main idea is an extension of the Simple Genetic Algorithm (SGA), which was called VEGA, and that differs from the first one only in the way the selection is performed. This operator was modified in such a way that after every generation a certain number of sub-populations are obtained. Hence, VEGA generates a set of possible solutions with different trade-offs among the objectives and this set of solutions is found using the Pareto dominance concept [143]. The basic principle states that a given solution x1 dominates another solution x2 if and only if:

- Solution x1 is not worse than solution x2 in any of the objectives; and
- Solution x1 is strictly better than solution x2 in at least one of the objectives.

As a consequence of its basic principle, VEGA-based algorithms generate solutions that are locally non-dominated, but not necessarily globally non-dominated. In fact, VEGA presents the so-called speciation problem (i.e., we could have the evolution of solutions within the population which excel on different objectives). Thus, for our problem with 3 objectives and a population size of M individuals, three

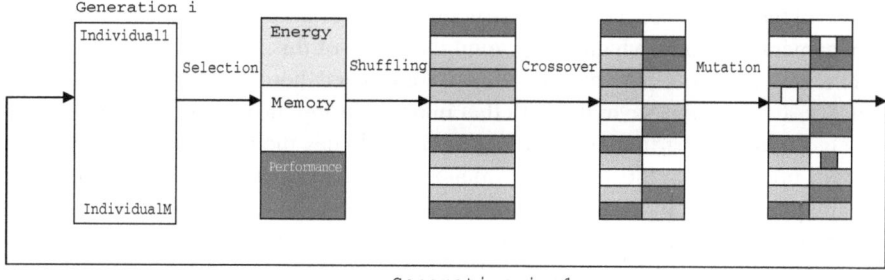

Fig. 4.21 VEGA-based design space exploration method

sub-populations of size M/3 each are generated. These sub-populations are mixed together to obtain a new population of size M, where we then apply the GA operators (crossover and mutation) to refine further the solution, as illustrated in Fig. 4.21. This process is repeated until no improvement occurs in any of the possible combinations generated in the last generation and in any of the target metrics. At this point, a Pareto curve of optimal solutions for the different optimization metrics can be generated (see Sect. 4.7 for some examples).

4.7 Application of the Exploration and Optimization Methodology to Multimedia and Communication Applications

In this section it is illustrated the application of the presented exploration and optimization methodology of DDT implementations using two industrial applications from the multimedia and network domains. The first case study is a 3D image reconstruction application (already described in Sect. 4.3) and the second one is a network scheduling application.

4.7.1 Exploration and Refinement of DDT Implementations in a 3D Image Reconstruction Application

The dynamic memory behavior of this application and its most important DDTs have already been described in Sect. 4.3. Apart from the software module that performs the reconstruction of the third dimension of the objects contained in subsequent frames [138], the complete application of the 3D image reconstruction system (see [160] includes several static data structures (mainly bi-dimensional arrays) to handle complex object textures, illumination analysis, colour interpolation in each portion of the image, etc. These data structures can be optimized using static memory management methods, which fall beyond the scope of this book.

The concrete software module that is optimized here has been presented in Sect. 4.3, and contains several DDTs that interact during the execution of its internal sub-algorithms. In particular, this software module has the most intensive dynamic memory utilization in the whole 3D image reconstruction system. Although it only contains approximately 10 K lines of code, it uses more than 35 % of the overall memory and energy resources of the system during the processing of each tested input stream of frames, which approximately contain in the range of 700–1000 frames. In the following paragraphs, the design and optimization methodology presented here is applied to optimize the implementation of the four main DDTs described in Sect. 4.3.

A first analysis of the run-time profiling information of the algorithm (see the left side of Fig. 4.23 and Sect. 4.3, we proceed with the optimization of the main DDTs of this algorithm, namely, CMCopyStatic (orginally implemented as a dynamic array), and CandidateMatches, ImageMatches and BestMatches, which are originally variations of double link list implementations. The evaluation of the optimal implementations for these DDTs is explored with the presented design flow based on multiple design objectives: memory footprint, memory accesses and energy consumption of the dynamic memory subsystem. As such, an *optimal solution* in the classical sense can normally not be identified. Instead, in the proposed methodology we use the concept of Pareto optimal points. A point is said to be Pareto optimal if it is not longer possible to improve upon one cost factor without worsening any other. As a result of the exploration process, Pareto optimal solutions are located for the 3D image reconstruction algorithm under study, based on normalized memory use, memory accesses and energy estimates (see Fig. 4.22) and the final solution depends on the restrictions of the designer and system. If constraints change (e.g., the available cycle budget), a new optimal Pareto point can be selected. By modifying the multi-objective exploration method using genetic algorithms (Sect. 4.6.3), more details about (sub-optimal) DDTs can be obtained, at the cost of an increased exploration time.

After the Pareto curve of solutions is generated, in the subsequent exploration step, our exploration tool suggested an ideal solution for the different DDTs according to the constraints entered. In this case, for illustration purpose we selected the minimal energy dissipation solutions and two-layered DDT structures, i.e., PAR(AR(Points)), with array sizes of 756, 1024, and 16384 bytes, for CandidateMatches, ImageMatches and BestMatches. As the second plot in Fig. 4.23 depicts, our DDT refinement methodology already attains a significant influence on performance and dynamic memory footprint.

Finally, because of the optimized DDTs found in the exploration, it is possible to remove CMCopyStatic applying algorithmic changes to the original implementation of the software module and refine further by combining the information it contains with the data stored in CandidateMatches. After this, the figures in Table 4.4 and the right plot in Fig. 4.23 are generated. They show that the accesses to ImageMatches are less than half of the original ones and the normalized memory use of CandidateMatches is 43.9 % less. The removal of CMCopyStatic

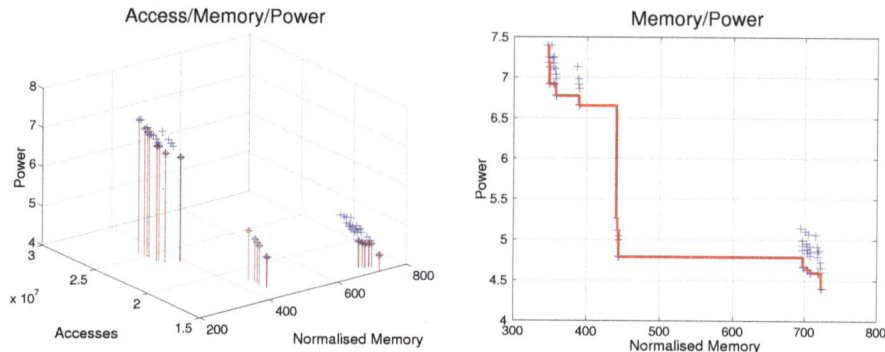

Fig. 4.22 The *left* figure shows a typical combination of Pareto optimal solutions. The global Pareto points are projected in the Memory/Accesses plain. They form a Pareto curve. The *left* shows a projection of this 3D space in the Power/Memory plane

Table 4.4 DDTs after DDTTR in the 3D image reconstruction system

Variable	Memory accesses	Memory footprint (B)	Energy 0.13 μm tech. (μJ)
ImageMatches	4.02×10^5	6.84×10^2	2.91×10^2
CandidateMatches	3.89×10^5	1.27×10^5	7.02×10^2
MultiMatches	7.68×10^3	3.81×10^3	0.03×10^1
BestMatches	7.16×10^3	3.78×10^3	0.02×10^1
Total	8.06×10^5	1.28×10^5	9.98×10^2

influences the normalized memory footprint (and removed the short memory peak used while copying), but has little effect on the performance (speed) of the overall program. This shows the importance of an appropriate selection of DDT implementations to speed up the application comparing the values of the X-axes from Fig. 4.23 to match two images (time values of $\times 10^7$ microseconds in the first plot, in the refined ones with our methodology only $\times 10^5$ microseconds). Furthermore, our proposed methodology to design and optimize the DDT implementation for each target embedded system and application reduces memory footprint and energy consumed of the DDTs, and enables additional global refinements.

4.7.2 Exploration and Refinement of DDT Implementations in a Network Scheduling Application

The second case study is a good representative from the communication domain, the *Deficit Round Robin* (DRR) application taken from the NetBench benchmarking suite [115]. It is a buffering and a fair scheduling algorithm that is commonly used for scheduling according to available bandwidth [147]. The algorithm is implemented in

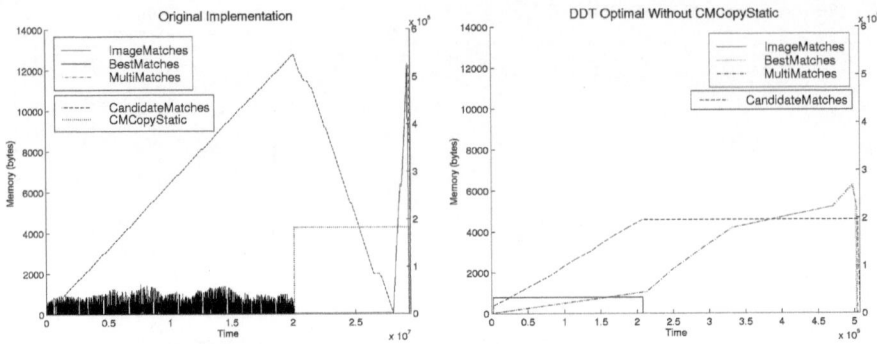

Fig. 4.23 Memory footprint over time. All plots mapped on the *left axis*, except CandidateMatches and CMCopyStatic (*right axis*). *Left* graph, original implementation. *Right* one, optimal implementation suggested by our exploration framework with CMCopyStatic removed

various switches currently available (e.g., Cisco 12,000 series). In the DRR algorithm, the scheduler visits each internal non-empty queue, increments a variable called deficit (representing the amount of data the queue can transmit) by the value quantum and determines the number of bytes in the packet at the head of the queue. If the variable deficit is less than the size of the packet at the head of the queue, then the scheduler moves on to service the next queue. If the size of the packet at the head of the queue is less than or equal to the variable deficit, then the variable deficit is reduced by the number of bytes in the packet and the packet is transmitted on the output port. The scheduler continues this procedure to the rest of the queues, traversing them in a round robin way.

The dominant data structures in DRR are two: the first is the class Packets, which is used to create and encapsulate the information of the packets to be scheduled in queues. The second one is the class Deficit_node, used to create the queue nodes, in which the packets are stored and scheduled.

In this case study, we simulate the application 500 times in order to explore the DDT combinations exhaustively (i.e., ×5 networks ×100 DDT combinations). The number of simulations using our methodology is reduced to 60. In each simulation 50, 000 packets were scheduled, and using five network configurations taken from five traces of network traffic from Dartmouth University [90]. After the exploration at application level, the pruning for two optimization objectives and the exploration at network configuration level, the simulation results were again automatically filtered to get the Pareto-optimal points depicted in Fig. 4.24.

In DRR in terms of energy, a 93 % reduction was achieved among Pareto-optimal points. The corresponding percentage of execution time reduction is 48 %, of memory footprint 53 % and of memory accesses 80 %. Note that these statistics come from comparisons only among the optimal DDTs. Other combinations can take more than double the time to run, 30 times more energy (consumed in the dynamic memory subsystem) and 4 times more memory accesses than the average of the Pareto-optimal.

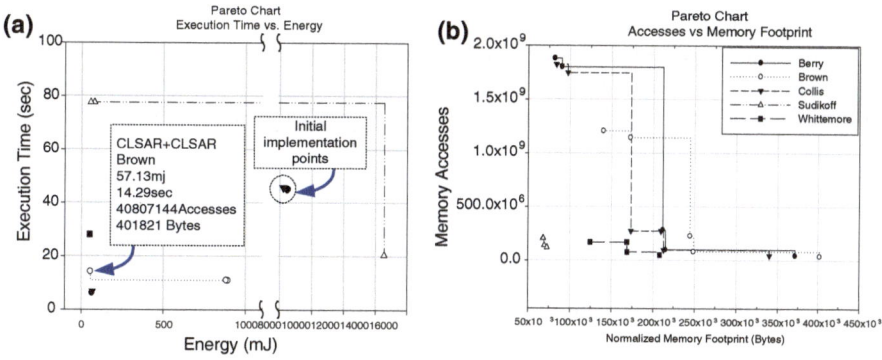

Fig. 4.24 DRR application simulated for 5 different networks (*curves*). Pareto charts for: **a** Execution time versus energy **b** Memory access versus memory footprint

79 combinations have been left out of the above statistics calculations due to their unacceptably high memory footprint (we considered 40 MBytes as a design constraint).

Finally, a comparison with the initial Netbench implementation is interesting (both data structures were initially implemented as single linked lists). The execution time is reduced by 22 % and the energy consumption is reduced by 80 %. The reason for the bad initial implementation is that the scheduling application creates huge lists of many packets and thus the memory footprint overhead of the 'next' pointers in each record of the linked list becomes very big. Most of our proposed DDT combinations (which are arrays or arrays of lists) do not have such a big memory footprint overhead and thus can be put in smaller memories (and smaller memories consume less energy per access).

4.8 Conclusions

New nomadic embedded devices have increased their capabilities and now complex multimedia and network applications can be ported to them. Such applications include intensive dynamic memory requirements that must be heavily optimized for an efficient mapping on embedded devices. To efficiently use dynamic memory in this applications, designers need to select suitable complex DDT implementations (dynamic arrays, linked lists, etc.) for the variables used in the running applications with respect to their specific embedded systems requirements (e.g., performance, memory footprint or energy consumption).

In this chapter we have presented a novel methodology and its supporting automated exploration that addresses the problem of implementing and designing DDTs for dynamic multimedia and network applications targeting nomadic embedded systems. This methodology shows that, the choice of an optimal implementation of a dynamic data type can be flexibly tuned to the specific needs of each application and particular set of constraints of each target embedded platform. Therefore, significant improvements in terms of energy consumption, execution time and memory footprint can be achieved, in combination with a reduced design time cost for DDT implementations.

In addition, we have presented an analysis of the dynamic behavior of multimedia and network application domains. We have also defined a set of systematic, high-level, data-structure transformations that enable the optimization of DDTs according to the specific design constraints of the target nomadic embedded system. Moreover, this methodology largely simplifies the exploration effort of suitable multi-layered DDT implementations for developers and construction of DDT implementations in an automated way through the use of a multi-objective optimization method based on evolutionary computation.

Finally, we have illustrated the application of the previous concepts of the proposed methodology to several case studies of multimedia and network applications of latest nomadic embedded applications. Our results with two industrial such applications, i.e., a 3D image reconstruction application and a Network scheduling application, show that the presented optimization approach can reach significant energy savings (80 % on average) and increase in performance (up to 40 %) of the original benchmarks implementation without augmenting the memory footprint consumed by the DDT implementations. Moreover, trade-offs among the Pareto-optimal choices provide alternative solutions to the designer.

Chapter 5
Intermediate Variable Removal from Dynamic Applications

Modern software applications for embedded systems have massive data storage and transfer needs. This is particularly true for applications that need to deliver a rich multimedia experience to the final user. This situation creates a bottleneck for the memory subsystem of an embedded system, which has severe memory footprint limitations due to size and cost factors. Additionally, the bandwidth of the interconnect is limited and can become overloaded by the memory accesses. Most importantly, the accesses to the memory affect the battery life of the device through the energy consumption of the physical memory (i.e., physical memories consume a certain amount of energy per access). Therefore, the memory accesses used by the data of the software application need to be minimized in order to meet the specification constraints of the embedded system's designer.

One of the main contributors to this energy consumption is the set of internal data structures that are present in these applications. The energy consumption of these data structures is due both to memory accesses to them by the application, as well as their memory footprint. Therefore, one of the main concerns [37] of the embedded designer is to reduce these accesses or the memory footprint to which they relate. These internal data structures are both of static (e.g., C arrays) as well as dynamic (e.g., vectors and lists in STL [145]) nature. While plenty of work is reported on reducing the memory accesses to static data structures [37], the work regarding dynamic data structures is more limited. One specific technique that has been employed to reduce the number of accesses to big memory structures, namely arrays, in the static context has been loop-fusion [110].

As an example of how loop fusion works for arrays, we present a simple example of two loops and show how these are fused together to remove the usage of the intermediary array a.

```
int a[N];
for (int i = 0; i < N; i++) {
  a[i] = f(i);
}
for (int i = 0; i < N; i++) {
  g(a[i]);
}
```

© Springer International Publishing Switzerland 2015
D. Atienza Alonso et al., *Dynamic Memory Management for Embedded Systems*,
DOI 10.1007/978-3-319-10572-7_5

It can be proven that if f has no data- or control- dependencies amongst calls or with g, then these two loops can be fused into the code shown below. It is clear that such a transformation has a potential for large gains in terms of memory bandwidth. First, the number of accesses are reduced as no longer a need exists to store and read the values from the array a, which depending on the size might be located in off-chip memory. Second, the memory footprint for a is completely removed, thereby reducing the total memory footprint.

```
for (int i = 0; i < N; i++) {
  g(f(i));
}
```

This technique has two limitations in the presence of dynamic data types such as dynamically allocated vectors or lists, from now on referred to as *sequences*. First, the access primitives for sequences are fundamentally different from those of arrays. While sequences support indexed-access, they are typically created with operations such as append, insert and remove through the use of iterators [145]. Second, the ability to fuse the above given loops into one loop depends on an affine mapping between the loop-index and the index into the array [140]. In the presence of sequence operations such as previously listed, this affine index is no longer compile-time manifest nor even compile-time decidable. More specifically, if we look at the following example, since it is not clear where each item falls in the underlying array of the vector, the above transformation no longer works.

```
vector<int> a;
for (int i = 0; i < N; i++) {
  if (h(i))
    vector.push_back(f(i));
}
for (vector<int>::iterator i = a.begin(); i != a.end(); ++i) {
  g(*i);
}
```

Nonetheless, with an ideal optimizer, this code could be turned into:

```
for (int i = 0; i < N; i++) {
  if (h(i)) {
    g(f(i));
  }
}
```

The approach presented in this chapter will include slightly more control-overhead, to handle more general cases than the one given above.

Three specific reasons exist why these intermediate sequences may be required from a developer's point of view:

- The control-flow structures of the production logic and consumption logic are not identical, thereby giving a mismatch between where elements would be produced and consumed. Manually fusing these loops would lead to a very complicated loop-structure.
- The code written for the production logic and the code written for the consumption logic are developed by two different developers and as such, the clearest interface between these two modules would be a sequence of elements.

- In general the manual fusing of production and consumption control-flow is a tedious and error-prone process. In languages where fusion [46] has been applied to sequences (e.g., Haskell), the effort is reduced as developers use functions that process whole lists at once, instead of manually writing loops. As such, fusion only needs to be done once in the library of functions that work on such sequences. Since in C++ each loop is different, this would induce a very high overhead if the developer wishes to remove intermediate sequences by hand.

Given these three points, a systematic method for fusing loops based on sequences is highly desirable as a pre-compiler step.

In this chapter, we first review in Sect. 5.1 the state of the art for intermediate variable removal in multimedia applications, where loop-fusion is the most important methodology where most of the existing approaches build upon. Then, in Sect. 5.2 we explore an alternative way to the classic loop-fusion techniques specifically targeted towards the use of these linear dynamic data types or sequences. As a first step towards the solution, we go back to the problem definition and explore what the fundamental operations of these sequences are. Based upon this information, the abstraction level at which this optimization is best performed, is formulated. The method (see Sect. 5.3) for this optimization will borrow techniques [69, 170] from the functional programming languages world to transform the production loops of these sequences into lazy generator loops, also known as *stream-generators*. Additionally, we formulate the constraints that give a conservative lower-bound on the applicability of this method. Finally, in Sect. 5.4 we show the results of such a transformation on several benchmarks and state some conclusions regarding the method.

5.1 Related Work

The current state of the art indicates that a large body of research to find solutions for scalar intermediate variable removal exists [126]. Additionally, De Greef [53] looks at ways to reduce indexed arrays accessed within manifest loop nests to their minimum necessary size. In [140] a similar model, known as the *Polyhedral model*, is used. Both approaches depend on compile-time analyzable for-loops with (piecewise) linear dependencies on the for-loop iterators in the addressing of arrays. More recently, work has been performed on inter-array analysis [165], which focuses on copy-propagation and employs the Polyhedral model. This framework employs sets of equations with linear affine indices and then utilizes matrix manipulations to enable transformations. All the numbers used in the expressions must be fully-manifest, and the method cannot cope with extended data dependent behavior. In addition, this work does not look at removing copies that are not pure copies of larger arrays. Finally, work was performed on stream like applications [3], however this work also requires that the number of elements being produced in each phase is linearly related to the number of elements consumed. Therefore, although the work also streams data consumption and production, it can only perform this optimization for linear state space systems, and not for non-manifest run-time dependent data consumption and production.

Within the literature of functional programming languages, a similar approach has been studied under the term *deforestation* [69, 170]. This approach is specific to pure functional programming languages such as Haskell [129]. Their compiler tries to remove intermediate data that is produced between two different functions to lower the data access overhead between functions that are chained together. An advanced form of deforestation is known as *fusion* [164], which targets also higher-order functions. Both deforestation as well as fusion are specifically targeted towards pure functional languages where the basic data type is the list and side-effects are not present. However, this cannot be directly reused within the context of object-oriented software applications due to the fact that the way data types are accessed there is much more complex than the way lists are accessible in functional languages and because *equational reasoning* [142] is not possible for non-pure languages (that is, languages with side effects).

Data management and data optimizations at design time for embedded applications have been extensively studied in the related literature [85]. [24, 132] are good overviews about the vast range of proposed techniques to improve memory footprint and decrease energy consumption in statically allocated data. Finally, from the method viewpoint, several approaches have been proposed to tackle this issue at the different levels of abstraction (e.g., memory hierarchies), such as the Data Transfer and Storage Exploration (DTSE) method [37]. However, all these approaches focus only at optimizations of global and stack data, which is allocated at compile-time. In this chapter, optimizations for dynamic heap data, which is allocated at run-time, are proposed. Obviously, the aforementioned approaches are 100 % compatible with the approach presented here and can complement it in order to optimize data of embedded applications, which is allocated both at compile-time and at run-time.

In summary, so far no research focuses on developing a systematic approach to remove *dynamic* data types from software applications with *non-manifest* and *data-dependent* behavior. Moreover, research has to be done on systematically eliminating variables that are not pure copies of global or larger dynamic variables, but also injective control-flow or processing relationships. Related, manual experiments were performed on such applications [97] focusing on reducing memory footprints, but no attempt was made yet to formalize the process in a systematic method, which also targets memory accesses. The main limitation in the previous methods with respect to dynamic data types is that the link between array-position and iterator is no longer manifest nor piece-wise affine. Even though the actual *iterator* used is linearly related to the for-loop instance in which it executes, due to the semantics of sequences, the link between this iterator and the actual storage inside the sequence is non-manifest and in most cases data-dependent. This is due to the fact that at any time an element is *added* or *removed* to a sequence (two operations that an array does not support), the index-to-value mapping of all the other subsequent elements are changed.

5.2 Problem Formalization

The objective of the method presented in this chapter is to remove intermediate sequences present in dynamic embedded applications. In this section, the meaning of *intermediate sequence* is defined. Then, the most appropriate abstraction level for the method presented in this chapter is explored. Finally, the problem of removing intermediate sequences is exposed.

5.2.1 Sequences

The method presented in this chapter targets applications that produce and consume `sequences`. In the actual implementation of the method we have specifically looked at STL sequences, however the idea is applicable for any sequence library that is similar to the ones in use of STL. In fact, the method is applicable in all other languages, as long as the data types that are being removed are similar to the type of sequences we have in mind. Therefore, it is important to define what we understand under the term `sequence` and what it means to be similar to the one we use. For this, we define the minimal specification to which the data type must comply for it to be removable by our method. Before this, we first introduce in a more informal manner what a sequence is, to give the reader familiarity with them, then we specify the exact expected behavior for them.

A sequence is a specific type of dynamic data-structure, namely a variable-sized container whose *elements* are arranged in a *strict linear order* [145] and which supports insertion and removal of these elements. Besides the ability to insert and remove elements, typical implementations of sequences (STL [145], Java Collections Framework [78],...) offer *iterators* to traverse through the elements. In the specific implementation of our method, we have opted to use the STL-variant of the interface due to the fact that most embedded software is written in C/C++. The interface of the STL sequence data types (list and vector) can be seen below.

- **constructor**—Constructs a sequence that is completely empty.
- **destructor**—Destructs a sequence as well as any elements it may contain, rendering all iterators to it invalid.
- **copy-constructor**—Copy-constructs a sequence that has a copy of the elements (sequences do not share data).
- **operator=**—Assignment operator, assigns a copy of the elements of one sequence to another, thereby destroying the former elements in the sequence assigned to. Just as the destructor, this makes any iterators of the sequence assigned to, obsolete.

- **begin**—Gets an iterator pointing to the first element of a sequence.
- **end**—Gets an iterator pointing to just past the last element of a sequence.
- **push_back**—Appends an element to the end of a sequence (Referred to as `append` from now on).
- **push_front**—Prepends an element at the beginning of a sequence.
- **pop_front**—Removes the front element of a sequence.
- **remove**—Removes an element from a sequence at the position indicated by the given iterator.
- **insert**—Inserts an element in a sequence at the position indicated by the given iterator.
- **increment**—Increments an iterator (`operator++` in C++) to point to the element strictly after the one it was currently pointing at.
- **dereference**—Gets the element pointed to by an iterator (`operator*` in C++).
- **set**—Replaces the element pointed to by an iterator (non-constant `operator*` in C++) with a new element.

STL supports a wider variety of operators, and these can be classified as shown in Fig. 5.1. In the case of random-accesses, we believe that it is more appropriate to use arrays instead of sequences and then employ optimizations as presented in [1, 140]. As for bidirectional sequential access, we leave this for future work and focus on unidirectional sequential accesses, which we believe to cover a major use-case of STL-like sequences. Other dynamic data types exist, such as sets and maps, but these fall outside of the context of this chapter.

We now formalize the exact properties that are required for data types for them to be considered as *sequence* within the context of this method. First of all, a sequence can be specified as a mapping from natural numbers to values: *Seq element = Integer > element*. Here we define the sequence-type (*Seq*) for some element-type (*element*) to be equal to a mapping from *Integer* to the element-type *element*. Before we continue, we state one of the first assumptions that we make in regards

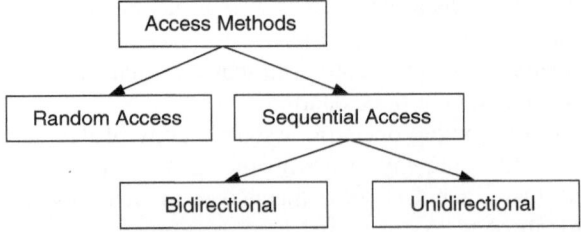

Fig. 5.1 Access method classification for sequences

to the element-type. Due to the fact that the basic operations as given above modify the index to element mapping, sequences cannot be treated as arrays. First of all, the addition and removal operations change the size of the sequence and therefore it is not possible to define the domain of a sequence for a loop. Secondly, each operation modifies the index to element mapping, therefore the mapping from loop-index in the production loop to sequence-index is not an affine mapping and may not even be statically analyzable.

Because we remove the permanent storage or elements and change the call-order of their constructors and destructors, constructors and destructors of the element-type *element* of any sequence cannot have observable side-effects. More specifically, constructors and destructor cannot have control or data dependencies that conflict with the rest of the code. Although actual allocations and de-allocations do form a unique behavior if in a certain order, we treat the allocation system as a black-box and treat each allocation call as if it is independent of other allocation calls. While internally this is not the case, we can nonetheless treat an allocation system as such as its behavior, from an abstract point of view is to return blocks of the requested size, where the actual address of these blocks is irrelevant.

The above assumption has one specific impact that should be noted. A lot of research has been performed in the optimization of dynamic memory allocation [14, 107]. And in that research, the actual behavior of the application in terms of allocation calls is relevant since the aim of these optimizations is to map data that is frequently allocated in memories closer to the processor. As such, since we change the allocation pattern of the application, it is important that those optimizations are performed after our optimization. Since we reduce the overall accesses of the application to big structures in general, it is beneficial to perform it before the dynamic allocation optimizations try to map the remaining blocks to the proper memory elements in the memory hierarchy. This introduces a dependence and ordering constraint in the overall design flow.

5.2.2 Intermediate Sequence Removal

To remove a sequence means to remove the elements of that sequence. Conceptually, it is impossible to remove the data stored in the elements, unless this data is never accessed, as it is part of the data-flow graph, however by shortening the lifetime of the individual data-elements, it is possible to ensure that the lifetimes of the individual elements are no longer overlapping. In this case, the footprint of the elements can be reduced, thereby decreasing the footprint, potentially to the size of a single element. We have specifically looked at multimedia software where often data is first produced in some production-logic and then consumed in a consumption-logic. Typically when large, dynamically varying quantities, of data are required, sequences are used to store these elements between the production control-flow and the consumption control-flow, to keep the different control-flows separated and understandable.

An example is the collisions that are detected between different points of two meshes that are then used to determine the resulting force on these two objects. Another example is the generation of salient features in an image to be used to match it with another image to do 3D extrusion, or even the matching of this set of points with the points in the other image that are then to be filtered for those candidates that fulfill a certain threshold.

The overhead that these intermediate sequences incur is both in the storage, and thus memory footprint, of these elements, as well as the memory accesses to copy the elements into the sequence and then read them back out of the sequence. Therefore, the removal of these sequences can lead to the reduction in both memory footprint as well as memory accesses and therefore energy consumption.

As such, the presence of sequences leads to a more modular and effort-less design of algorithms. Therefore, we propose in this chapter a systematic approach to remove the intermediate sequences in the final phase, namely as a precompilation step. To remove these intermediate sequences, two possible approaches exist for reducing the lifetime of the data stored within these sequences. Either the consumption logic is moved closer to the production logic, or the production logic is moved closer to the consumption logic. The former has typically been performed by inlining the actual consumption of the individual elements into the control-flow of the production loop [97], however this is only applicable if the control-flow of the consumption logic is very trivial or is parallel to the production logic control-flow. The latter can be performed using a similar transformation, or the control-flow can be retained but be delayed such that it is only invoked when an element is required. Certain languages already support such mechanisms, for instance streams or generators in python.

Through the use of these streams we are in essence turning the production loop into a separate thread and our transformation rules will ensure an idealistic scheduling between the consumer and producer thread such that the producer thread is only invoked when an element is required by the consumer loop. More specifically, streams employ the use of green, or cooperative, threads such that the overhead of this thread-switch is only the overhead of a function call. In fact, the consumer loop will invoke the producer thread whenever an element is required. And since this is done through cooperative threads, this amounts to calling a function that calls back into the producer code at the point where it last stopped and returned an element.

Two ways exist to achieve the streaming of production-loops.

- Provide a library of stream-primitives that allows people to write streaming production loops. While this seems an attractive option, the downside of this is that standard concepts like "for" and such cannot be reused and would have to be rewritten in some form of a streaming loop. Additionally, this would require that the application developer learns a new way of working. Finally, it means that the technique cannot be employed on existing legacy software without a rather in-depth rewrite.
- Provide for a mechanism that is able to pause the production-logic as well as automatically rewrites typical sequence operations to enable this pausing and passing out of an element. We have opted for this second approach because it is more

generally applicable and has a lower-effort requirement from the developer's standpoint. Note that this still uses some primitives, however these are automatically generated from the code that uses sequences. As such, the developer's effort is reduced drastically.

As such, we have reduced the problem to two smaller ones: We need a mechanism that is able to pause control-flow or basically allow for interleaving at run-time of the production and consumption control-flow; and we need to rewrite rules that transform typical sequence operations to employ this mechanism. In this chapter we present a method to tackle both these problems. As will be shown in Sect. 5.3.1, the patterns that are encountered in C++ code that uses STL can lead to code that does not lead to pure streams. More specifically, because sequences are not purely concatenated, due to the existence of the insert and remove operation, the rewriting of sequences to streams is non-trivial. This is another reason why the purely library-based approach would be less trivial for developers to use.

5.2.3 Abstraction Level

As final aspect, it is important to study at which level of abstraction the problem is to be solved. We have already alluded to this in the previous section. More specifically, one can envision three levels of abstraction. These level of abstractions relate to the primitives that the method has to deal with.

- Loop level—At this level of abstraction, the primitives are the loops that consume and produce sequences. More formally they are referred to as the cata-, ana- and hylo-morphisms [114]. This is the approach that is taken by the Fusion work in Haskell [46]. In such languages, due to the leveraging of first-class functions and higher-order functions, only a limited set of primitives are defined that work on complete lists. Due to the way that C++ programs are written, this is not a viable option as many programs have explicit loops and each such loop would introduce a new primitive.
- Data Type Interface level—At this level of abstraction, the primitives are the different operations provided by the sequence interface. We have opted to go for this level of abstraction as the interface that is defined for different implementations of sequences is rather standardized (i.e., STL [145]). Additionally, the number of primitives defined at this level of abstraction are a limited set. Moreover, this is the level of interface that developers who write code using sequences adopt, as sequences are a well defined library.
- Pointer level—This is the lowest level of abstraction, where the interface of the sequence data type is actually broken open and the internals are looked at. It is our belief that this abstraction level is too low for this method. First of all, at this level the use of pointers is prevalent which can quickly lead to aliasing-analysis [93] which is a hard problem. Secondly, widely varying implementations of sequences

exist where the internals look completely different. As such, the fact that it is a sequence that is being dealt with would be lost in the noise.

5.3 Method

In the previous section we explored *why* IVR is a relevant problem and *what* specific problems are required to be tackled to enable the proposed optimization. In this section we first expose *which* approach is taken to tackle the above problems, and then we show *how* to implement this method.

To enable our proposed approach, we need to tackle two specific aspects. First, we need a set of transformation rules that rewrite the sequence operations to operations that employ the green-thread primitives to alter the scheduling of the consumption and production control-flow at run-time. Secondly, we need a way to exline [163] the production control-flow so that it becomes its own cooperative thread. This will lead to specific constraints in terms of the usage of this method. Since the production control-flow scheduling is now changed with respect to the original application, we need to ascertain that no read-write or write-write conflicts are present that are introduced by this method. An overview of this method can be seen in Fig. 5.2. More specifically, the method consists of the following six steps:

1. First, the application is profiled to determine the dominant sequences. This is done by seeing which sequences have many accesses and take up a lot of memory footprint, and are produced first and then consumed, instead of having production and consumption to these sequences happen in an interleaved fashion.
2. Then, the constructor and destructors of the elements belonging to these sequences are removed, if necessary. This is to ensure that we can abide by the constraints

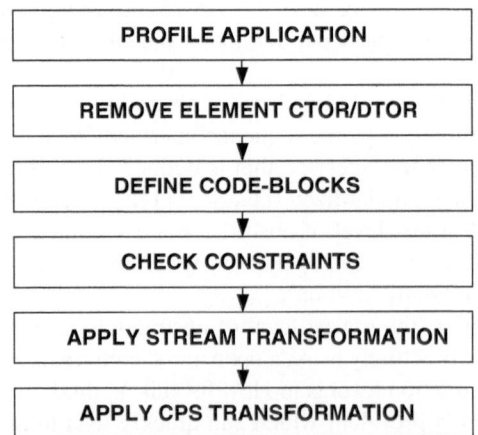

Fig. 5.2 Overview of the method

formulated in Sect. 5.2.1. The details of such a transformation are outside the scope of this book and belong to future work. If the constructor and destructor have no observable side-effects on global state, then this can be as simple as inlining the code in place. Additionally, we check whether the sequence in question is used unidirectionally, to ensure that our method will work for the targeted sequences.

3. Next, the code-blocks of the production code of these sequences are determined following the rules of Sect. 5.3.1.
4. Then, it is determined whether these code-blocks do not break the constraints as defined in Sect. 5.3.4.
5. Next, the sequence production loops are transformed to stream generators as detailed in Sect. 5.3.1.
6. Finally, the control-flow primitives that are used for these stream generators are removed using the transformation detailed in Sect. 5.3.2.4.

As mentioned in Sect. 5.2.2, two specific problems need to be tackled. Through the above steps, which tackle these two problems, we in essence turn production-control-flow with their associated sequences into stream-generators. What this implies is that we have an idealistic scheduling between the production and consumption loop. This idea is the same as applied by more traditional loop-fusion techniques, although the difference is that in this case, the scheduling is performed at run-time through the use of cooperative threads. In traditional loop-fusion techniques, the scheduling is statically analyzable due to the regularity of the loops and data-accesses. Because the access operations to the sequence in terms of the accessed index can no longer be correlated to the loop-index, such a static approach is no longer possible and the scheduling of the production and consumption accesses is delayed until run-time.

The cooperative green-thread primitives are based on a limited version of *delimited continuations* [39] (see also Sect. 5.1). Continuations are reified values of the current control-flow, and thus allow one to store this in a variable before jumping to another part of the control-flow. By making it storable, it is possible to later reinvoke this paused control-flow to continue where it left off. Continuations are not a novel subject and have been treated extensively in literature [48, 49, 65], however the introduction of them in embedded software is novel. As such, we will first introduce transformation rules for rewriting sequence operations to become stream operations in Sect. 5.3.1. Then, in Sect. 5.3.2 the primitives that are used for these transformation rules as well as a transformation to remove them and return back to plain C++ code are presented. Finally, in Sect. 5.3.4, we specify the constraints that the transformation must respect in order to ensure that no read-write or write-write conflicts are introduced. Besides just access constraints, this section will also cover the extra challenges that are introduced by the fact that we are dealing with a stack-language instead of a garbage-collected language, and therefore, where variables and data have a bounded lifetime.

As mentioned in Sect. 5.2.1, sequences support various different operations, and therefore, to maximize the applicability of the method presented in this chapter, we must ensure to cover as many of these operations as possible. Just like optimizations that focus on static manifest array code, the optimizations here focus not only on specific operations but also the control-flow that surrounds them. As such, the concept of an iterator exists. However, instead of dealing with loop iterators, these iterators are embodied into actual abstract entities, namely the sequence iterators. Additionally, since these iterators are opaque, we cannot perform the optimization completely at compile time, which is why delimited continuations are used to schedule the proper order of operations at run-time. However, it does mean that like static manifest approaches, we must deal with transformations that target more than just the specific pieces of code that use the sequence operations, but also look at the control-flow that surrounds these pieces of code.

One advantage of performing the scheduling at run-time instead of compile-time means that we can deal with more types of control-flow (for instance while-loops or predicated statements). The only limitation to what control-flows can be handled resides in what control-flows we can 'invert', which is what Sect. 5.3.2 focuses on. Due to the fact that we can handle more than nested-for loops, the usage of the terminology *loop-nest* as used in more traditional literature is less than apt. Therefore, we introduce a new term to refer to segments of the control-flow, namely *code-block*. In this chapter we assume only intra-procedural code-blocks, leaving inter-procedural code-blocks as an extension.

5.3.1 Rewrite Rules

In this section we introduce the rewrite-rules that are used to transform sequence operations into operations that rely on the delimited continuation primitives. In [139], we have presented a subset of these rules that were based on pattern-matching whole control-flows. By introducing the concept of delimited continuations and the ability to pause computation, we can now just rewrite the primitives that are used, thereby enabling many more control-flows to be handled.

One of the main limitations in the method is that unless otherwise stated, we do not allow the mixing of different sequence operations within one code-block. Although this may seem like a major limitation, if the limitations of vectors as they exist in common libraries are studied, then this will turn out not to be such a limitation. Specifically, the use of any sequence operation that changes the number of elements in it tends to invalidate the iterator. The operations that do operate on iterators (`remove` and `insert`) return a new iterator pointing at the proper place but with a valid iterator. The reason that this is done is that iterators are actually glorified pointers and insertion or removal will shift elements or even cause memory re-allocation, thereby invalidating old pointers. Since we want to define code-blocks to be the smallest possible code-block that is still meaningful to transform, we only allow code-blocks to pertain to the use of one iterator, which in the case of `append` is an implicit end iterator. From this it becomes obvious that it is impossible to use `append` and an iterator-based operation in the same code-block. On the other hand, it is possible to have both `insert` and `remove` operations inside one code-block as long as they're based on one iterator. Since each code-block only deals with one iterator, be it explicit or implicit, we will refer to it as the code-block's iterator.

Each code-block along with the iterator is transformed to a stream-transformer that is layered on top of the previous stream-generator. The first stream-transformer is layered on top of a stream-generator that generates 0 elements, namely the stream-generator equivalent of an empty sequence. So in essence, a stream-transformer has as type $stream - generator - > stream - generator$. On top of the final stream-generator, we add an iterator interface along with a mockup sequence interface such that the consumption of the sequence can still occur as if it is an actual sequence instead of a stream.

The second limitation is that a sequence operation of a sequence that is transformed with this technique must map exactly to one code-block. More specifically, this means that all operations must belong to a code-block, meaning that no operations may remain that are still on the original sequence, and secondly that an operation may not belong to more than one code-block. The latter means that a code-block may not reside within a loop. While extensions could be possible to allow this, we do not believe this to be a good idea. Each code-block imposes an extra layer in the stream-generator and incurs a slight amount of memory due to the stack-saving of the code that will be put in the continuation belonging to that code-block. In the extreme case, given a for loop that performs a set of appends, one could make a code-block for each append iteration instead of for that statement, and this would mean that as many code-block layers are present as original elements. Since typically a code-block will incur more memory overhead than a single element, this is obviously a loss and not a gain in memory footprint.

The application of the rewrite rules consists of a sequence of steps:

1. The scope of the code-block is defined. The definition of this scope is defined per rule and is subject to extra constraints formulated in Sect. 5.3.4.
2. This scope is delimited with the **reset** construct that will be described in Sect. 5.3.2.
3. The operations inside the code-block are transformed to operations based on continuations, based on the rewrite-rules. These rules are specific per operation being transformed and are detailed below.
4. Based upon the specific rewrite-rule, this transformed code-block is wrapped in some logic, referred to as the *thunk*, to create a stream-transformer.
5. Finally, the different stream-transformers are stacked on top of an empty stream-generator and are wrapped in an iterator interface.

All the thunks, and by extension all the stream-generators, share a common interface. Specifically, it consists of two operation (where the element type of the original sequence, and thus the stream-generator, is E):

- a constructor that takes the stream-generator before it and creates a new stream-generator; and
- a method `bool next(E & e)`, which generates the next element and stores it in e and returns true, if no next element exists, meaning the stream has finished, it returns false and e is not modified.

We now detail the specific rewrite-rules, along with their scope-definition and the logic to create stream-transformers. These are the two specific parts, namely step four and five of this part of the method, and are specific per operation being transformed. Note that since we do not allow different iterators to be used in one production block the patterns below cover all the major usage cases for STL unidirectional sequences.

- **append**: In this case, the iterator is implicit, and the start of the code-block is defined to be *at least* before the first append and ends *at least* after the last append. This is of course subject to the constraints given above, namely that operations on different iterators may not be within the same code-block. It is important to note that the terminology used here is *at least* and not *right*, we refer the reader to Sect. 5.3.4 for further details. Because the iterator is implicit, it is possible to have a code-block defined as two code-blocks, and this can be useful when taking into account the constraints in Sect. 5.3.4. However, all things being equal, it is best to minimize the code-blocks for the reasons given above. The transformation rule for append is rather straight-forward. Since appends add elements in the same order that they are consumed, we transform append(e) to a co-routine **yield**(e) statement, that will pause the loop generating the data and simply return the element. The thunk is shown below. Specifically, it will first try to consume all the elements in the previous generator, if those are finished, only then will it start returning its own elements. Obviously, the generator is only finished after both the previous generator and the current continuation are finished. Note that this is actual C++ code and not pseudo-code.

```
template<typename E, typename Continuation>
class implicit_generator : public Generator<E> {
  typedef Generator<E> Prev;
  public:
    implicit_generator(Prev & prev, Continuation continuation)
      : previous(prev), continuation(continuation) {}
    bool next(E & e) {
      if (previous.next(e)) return true;
      else return continuation.invoke(e);
    }
  private:
    Prev & previous;
    Continuation continuation;
};
```

To demonstrate this transformation, we use the following piece of code:

```
typedef vector<int> DataSequence;
DataSequence data;
for (int i = 0; i < n; ++i)
  if (foo(a[i]))
    // STL equivalent for the concept of append
      data.push_back(i);
```

After the transformation, the code would look as follows. Since this loop uses the **append** primitive, the generator used is the implicit_generator and the append statement simply becomes a **yield** operation. The entire loop construct is now wrapped in a **reset** construct that specifies where later control flow transformations will be required as specified in Sect. 5.3.2. This is required, since the **yield** statement is basically a way to stop the co-routine specified by the reset block, temporarily pausing the execution of that block to return a value. Note that since we started with an empty sequence prior to this loop, the previous generator used to generate any data is simply the `empty` generator which simply always returns false when asked for the next element. This is the same one as the one shown below, namely the `clear_generator`.

```
typedef stream_vector<int> DataSequence;
DataThunk data_thunk = reset {
  for (int i = 0; i < n; ++i)
    if (foo(a[i]))
      yield(i);
}
Generator<int> empty;
implicit_generator<int, DataThunk> data_generator(
  empty, data_thunk);
```

- **insert** or **remove** : In this case, where an explicit iterator is used, the start of the code-block is defined to be *at least* before where the iterator is set to the `begin` of the sequence in question. The end of the code-block is defined to be *at least* after where the iterator is used last. The transformation rule is as follows:

  – Each **insert**(i,e) is transformed to **yield**(PUSH,e)
  – Each **remove**(i) is transformed to **yield**(POP)
  – Each **++i** is transformed to **yield**(ADVANCE)

The thunk for the code-block is detailed below. Specifically, if the continuation is finished and no more actions are to come from it, we can just empty our stack and then finish with the remaining elements in the previous stream. On the other hand, if the continuation returns a PUSH, then we simply put that value on the stack as a value that should come after the current element and keep looking for the current element. If the continuation returns a POP, this mean that the original code removed an element. Therefore, we either check the stack of inserted elements and remove one. If no elements were inserted, then this means that we need to remove the next element from the underlying stream and thus call it next, tossing away its value, and continue looking for the next value to generate. We do not consider the case that the previous stream is empty, because this means an error. Finally, whenever the continuation returns an ADVANCE, this means that no longer any insertions or removals will be performed at the current position, which means that we can return the element, either the last one that was inserted (and thus on the stack), or an element from the previous generator. If the previous generator returns the fact that it's empty, then this means that the original code was incorrect, since this means that the iterator was pointing past `end()`. To make it symmetrical with the finished continuation, we simply then refer to `false`, but give no guarantees about correctness (especially when the generator is then reinvoked later).

```
enum Operation {POP, PUSH, ADVANCE};
template<typename E, typename Continuation>
class explicit_generator : public Generator<E> {
  typedef Generator<E> Prev;
  public:
    explicit_generator(Prev & prev, Continuation continuation)
    : previous(prev), continuation(continuation) {}
    bool next(E & e) {
```

```
              Operation op;
              E temp;
              if (continuation.invoke(op, temp)) {
                switch (op) {
                  case POP:
                    if (stack.empty()) { previous.next(temp); }
                    else               { stack.pop_back();    }
                    return this->next(e);
                  case PUSH:
                    stack.push_back(temp);
                    return this->next(e);
                  case ADVANCE:
                    if (stack.empty()) {
                      if (previous.next(e)) { return true;  }
                      else                  { return false; }
                    } else {
                      e = stack.back();
                      stack.pop_back();
                      return true;
                    }
                }
              } else {
                if (stack.empty()) {
                  if (previous.next(e)) { return true;  }
                  else                  { return false; }
                } else {
                  e = stack.back();
                  stack.pop_back();
                  return true;
                }
              }
            }
          private:
            Prev & previous;
            Continuation continuation;
            vector<E> stack;
        };
```

To demonstrate this transformation, we use the following piece of code:

```
  typedef vector<int> DataSequence;
  DataSequence data;
  iterator it = data.begin();
  for (int i = 0; i < n; ++i)
    if (foo(a[i])) {
      it = data.insert(it, i);
      ++it;
    }
  iterator it = data.begin();
  for (int i = 0; i < n; ++i)
    if (bar(a[i])) {
      it = data.remove(it, i);
      ++it;
    }
```

What is important to note here, is that STL requires that iterators can be invalidated when an insert is used. As such, we define the block that we want to tackle as the block defined by the first definition of the iterator using the **begin** statement, and then until the last use of an iterator that is defined by the return operation of either an **insert** statement or a **remove** statement. Note that the requirement here is that each iterator is only used once, and then a new one is returned when it is consumed by a **remove** or **insert**. So in the code sample above, two blocks are defined. The transformation will result in:

```
typedef stream_vector<int> DataSequence;
DataThunk data_thunk = reset {
  for (int i = 0; i < n; ++i)
    if (foo(a[i])) {
      yield(PUSH, i);
      yield(ADVANCE);
    }
}
Generator<int> empty;
explicit_generator<int, DataThunk> insert_generator(
  empty, data_thunk);
DataThunk2 data_thunk2 = reset {
  for (int i = 0; i < n; ++i)
    if (bar(a[i])) {
      yield(POP);
      yield(ADVANCE);
    }
}
explicit_generator<int, DataThunk2> remove_generator(
  insert_generator, data_thunk2);
```

What is important to note here is that each block of code results in a new different class, since once the code inside **reset** block is transformed, it will result in a new class definition containing the required state and the code of the control-flow transformation.

- **clear**: The code-block of a clear operation is exactly that statement. The stream-transformer should obviously be something that ignores the stream-generator below and always says the stream is empty. The thunk for this operation is trivial, and in fact since the code-block contains only the single statement clear, we do not even need to use a continuation.

```
template<typename E>
class clear_generator : public Generator<E> {
  typedef Generator<E> Prev;
  public:
    clear_generator(Prev & prev) {};
    bool next(E & e) { return false; }
};
```

The last aspect that is required is an object to wrap the full stream-generator in that will conform to a (read-only) iterator, as well as faked sequence implementation that allows for the operations begin and end. We do not detail the aspects of the fake sequence, as this is rather straightforward, and instead will illustrate it later in an example in Sect. 5.3.3. The end iterator will just return a mock iterator for which finished is set to true. The begin iterator will return an iterator as shown in Fig. 5.3. Besides this, also a top-level stream-generator is required that basically generates no elements.

5.3.2 Delimited Continuations

In this section we introduce what delimited continuations are to give the reader a feeling for what they do specifically. Delimited continuations enable one to mark, or *delimit*, a region of the control-flow that is labeled with a *reset* point. Then, one can

```
template<typename E>
class Generator {
  public:
    bool next(E & e) { return false; }
};




template<typename E>
class stream_iterator {
  typedef Generator<E> Gen;
  public:
    stream_iterator()
      : n(0), finished(true), generator(empty_generator<E>()) {
    }
    stream_iterator(Gen & gen)
      : n(0), finished(false), generator(gen) {
      finished = !generator.next(element);
    }
    bool operator==(const stream_iterator & other) const {
      if (finished && other.finished) return true;
      if (finished) return false;
      if (other.finished) return false;
      return (n == other.n);
    }
    bool operator!=(const stream_iterator & other) const {
      return !(*this == other);
    }
    stream_iterator & operator++() {
      if (!finished) {
        ++n;
        finished = !generator.next(element);
      }
      return *this;
    }
    const E & operator*() const { return element; }
  private:
    bool finished;
    E element;
    int n;
    Gen & generator;
};
```

Fig. 5.3 Implementation of the interface to wrap a stream-generator in an iterator

shift out from any point inside the control-flow that is delimited and capture all the processing that needs to occur until the end of that region and *reify* it. This reified captured continuation can then be saved, passed around or invoked. When invoked, computation will continue from where the shift point left the reset region.

5.3.2.1 Overview

As mentioned, to enable the method we need to pause and reify the production control-flow logic. By doing this, we can turn our production loops that use sequence operations into stream generators through the use of these pausing primitives as well as the rewriting rules that were be presented in Sect. 5.3.1. Certain imperative languages, such as Python, have the **yield** keyword. These are then used to create generator-functions that generate streams [144]. Although useful, this particular feature is not generic enough for our requirements. The pythonic **yield** keyword only works for entire functions, pausing the entire execution of the function. If we were to use such a feature, we would first need to exline the selected-region of code that needs to be paused. Additionally, as seen in Sect. 5.3.1, we will need primitives that enable more than just passing out a value. Finally, no typing rules are present for the pythonic **yield** as Python is a dynamically typed language. For C++ this would need to be re-examined to give it a proper type. Additionally, because C++ is a scoped language and has such concepts as constructors and destructors, these need to be looked at.

In the context of functional languages, a more fundamental primitive named *continuation* exists. In [89], it was shown that streams can be encoded based on top of *delimited continuations* [39], a particular type of continuations. We use a similar approach, though, as previously shown in Sect. 5.3.1 that the types of code-blocks we encounter are more than purely stream-based. Additionally, we provided automated rewrite rules from normal sequence operations to these stream-primitives. Finally, we introduce the concept of delimited continuations in C++. Several aspects are present here that are less than trivial. First of all, how delimited continuations jointly operate with the C++ typing system is not defined. Secondly, besides tricks such as setjmp and longjmp, no defined way exists for capturing the current state to reinvoke it later. Finally, continuations in functional languages have the same interface as any other function. This works well in a language that allows for first-class functions, but this is not the case of C++. To cure these problems, we introduce a new version of these concepts within the context of C++, define their meaning and rules, and finally define a mapping back to pure C++ without these constructs.

The observant reader might have noticed that we used the keyword **yield** in Sect. 5.3.1. This is a keyword that we will introduce now that differs somewhat from the pythonic version. First of all, this keyword will be properly typed, and secondly it will not be yield from a function but from a reset point. A few differences exist in the way that we implement delimited continuations with regards to how they are defined in scheme or other functional programming languages. These differences are due to two reasons. First, we want an efficient mechanism to enable this without inducing a lot of stack-copying overhead. Secondly, we want to make sure that this fits properly within the type-system of C++. Finally, due to the fact that we only need to be able to yield out of a continuation and then start where we left off, we do not need to fully reify the continuation. More specifically, in the typical case we only need to support one-shot continuations, more details regarding this are explained further below. Hence, we will refer to these continuations as *restricted delimited continuations*. The main difference between function languages and C++ is twofold:

- C++ creates a difference between expressions and statements where no such distinction exists in functional programming languages.
- C++ has scoped constructor and destructor behavior, while functional programming languages rely on garbage collection.

5.3.2.2 Syntactic Constructs

We now introduce the workings of this keyword, along with **reset**, as well as **invoke**. Although already referred to in Sect. 5.3.1, we now show an example use of the constructs to give the reader an intuitive feeling on how they work. An example of their workings can be seen below. In this case, a continuation is created through the use of the **reset**. This continuation is then invoked as long as it has not finished, passing it a reference where it should yield data to. The first time the continuation is invoked, it starts at the begin of the reset point, then when it comes across the **yield** operation, it stores the current state and returns **true** to signify the continuation is not finished yet, as well as yielding the value i into the passed a reference.

Continuations c = reset int i = 0; while (i < 3) std: :cout   Output: 0 foo0 1 foo1
« i « std: :endl; yield(i); i++;   int a; while (c.invoke(a))   2 foo2
std: :cout «"foo" « a « std: :endl;

5.3.2.3 Definition of the Typing Rules Language

We now introduce the type-rules related to the use of the continuation primitives. A continuation is parametrized by the types that may be yielded out of it. Since yield may yield multiple arguments at once, we used the following notation $\ll T_1, \ldots, T_n \gg$ to refer to a list of C++ types. For conciseness reasons, $T_1 :: \ll T_2, \ldots, T_N \gg$ is equivalent to $\ll T_1, \ldots, T_n \gg$. Except for the extra notes that follow, the format is the same as any other type-rule presentation, where Γ stands for a type-environment mapping variables to types, and a rule in the form of $\Gamma| - s : \tau$ stands for the fact that within the typing environment Γ, the statement s has a *yield-type* τ, where τ is a list of C++ types. We overload the same notation for the typing of expressions, where instead of a yield-type, the type of an expression is simply it's C++ rtype (where rtype for all types is simply the type, except for references of the form $T\&$, where the rtype is T). We do not introduce the typing rules for the extension of the Γ scope as these are rather straight forward and can be found in the relevant literature. Additionally, we assume standard C++ typing-rules for the expressions and leave these out for brevity. Finally, the type any statement not described is assumed to be $\ll\gg$. The type of a continuation is parametrized by a list of C++ types, denoted as $\ll T_1, \ldots, T_n \gg$. As a last note, syntactically, **yield** may only be used inside **reset** scopes.

$$\frac{\Gamma| - s : \tau}{\Gamma| - < while > s : \tau} \quad \frac{\Gamma| - s : \tau}{\Gamma| - < if > (e) \, s : \tau} \quad \frac{\Gamma| - s : \tau}{\Gamma| - \{s\} : \tau} \tag{5.1}$$

$$\frac{\Gamma| - s : \tau}{\Gamma| - < reset > \{s\} :< continuation > [\tau]}$$

$$\frac{\Gamma| - e_1 : T_1 \quad \ldots \quad \Gamma| - e_n : T_n}{\Gamma| - < yield > (e_1, \ldots, e_n); :\ll T_1, \ldots, T_n \gg} \tag{5.2}$$

$$\frac{\Gamma | - s_a : \tau_a \quad \Gamma | - s_b : \tau_b \quad join(\tau_a, \tau_b) \Downarrow \tau}{\Gamma | - s_a \; s_b : \tau} \tag{5.3}$$

$$\frac{\Gamma | - s_a : \tau_a \quad \Gamma | - s_b : \tau_b \quad join(\tau_a, \tau_b) \Downarrow \tau}{\Gamma |- \; < if > (e) \; s_a \; < else > s_b : \tau} \tag{5.4}$$

$$\frac{}{join(\ll \gg, \tau) \Downarrow \tau} \quad \frac{}{join(\tau, \ll \gg) \Downarrow \tau} \quad \frac{join(\tau_a, \tau_b) \Downarrow \tau \quad T_a \equiv T_b}{join(T_a :: \tau_a, T_b :: \tau_b) \Downarrow T :: \tau} \tag{5.5}$$

The type of a $< continuation >$ $[\ll T_1, \ldots, T_n \gg]$'s method **invoke** is given by the following C++ definition:

```
class continuation {
  public:
    bool invoke(T_1 & p_1, ..., T_n & p_n);
};
```

5.3.2.4 Transformation

Now that we have shown how these restricted delimited continuations can be used, we present how to compile them away to go back to C++, after having applied the rewrite rules of Sect. 5.3.1. We assume that the **reset** and **yield** constructs have been introduced correctly and leave further concerns of correctness in regards to the whole of the method for Sect. 5.3.4. More specifically, this is related to writes being delayed to a different point in time. However, assuming that the use of the **reset** block is written as intended, then this comes down to a proper transformation to code using these constructs, which is what Sect. 5.3.4 focuses on. To remove the constructs we introduced, a variety of ways to go are present. We list here the most obvious solutions along with the one we have picked and the motivations therefore. It should be noted that our method does not hinge on the specific choice we make, and therefore under other assumptions, the other options would still enable the method.

- Use the setjmp/longjmp constructs along with explicit stack-copying to save the point of the stack where we were when yielding. Although a possibility, this is a very ad hoc approach without a strong semantic basis.
- Write an explicit state-machine to encode where exactly we are in the **reset** scope. This approach would be ideal if our target language were something lower level than C++ (e.g., C). However, since we wish to remain at the C++ level after the transformation, this is a less interesting option as we would have to heap-allocate all variables.
- Perform a continuation-passing-transformation to the code in the **reset**. We have opted for this last choice as it allows to only heap-allocate the variables that need to survive across a **yield**.

We present the basis of the continuation-passing-transformation for a subset of C++ and leave the completion to full C++ as an exercise for the reader, because fundamentally the principle would be the same. First we introduce the subset of C++

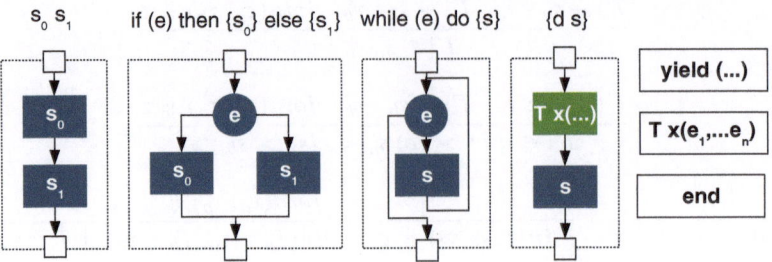

Fig. 5.4 Control flow constructs of *SIMPLE*, a subset of the C++ language

that we are interested in, named *SIMPLE*, noting that the only parts that our transformation is concerned with are scope-declarations and control-flow structures. This
language, nor any extension thereof, can support the use of **return** as it would
not be semantically clear where it should return from: the original function where
the continuation was captured or the continuation itself. We also do not present the
details for non local control-flow within a procedure (**break** and **continue**), however these are easily added, though they add some complexity to the second step as
scopes behave slightly different in the presence of these non-local constructs. The
syntactical constructs are shown below and presented visually in Fig. 5.4, where *act*
stands for a plain statement. The definition of variables has been distinguished with
a different color, as has the scope within which they reside.

$$s ::= \quad act|s_0\ s_1|\{d\ s\}| < yield > (e_1,\dots,e_n)| < finish > \tag{5.6}$$
$$| \quad < if > e < then > \{s_0\} < else > \{s_1\}| < while > e < do > \{s\} \tag{5.7}$$
$$d ::= \tau\ X(\overline{e}); \tag{5.8}$$

The transformation consists of four steps.

- In the first step, the block of code that is contained inside the **reset** point is hoisted
  out and placed apart for transformation. We add another syntactical construct
  (**finish**) to keep track of where the end of the **reset** point was. Additionally,
  each yield statement is tagged with a unique-number. Similarly, each **while**
  and **if** is also uniquely annotated in a similar manner, thereby creating unique
  identifiers for fix-points and splits in the control-flow. The original code with the
  reset block is put apart for transformation, and the reset block assignment to the
  continuation object turns into a plain constructor call, passing in the variables that
  are declared outside of the reset-point. A skeleton of what the class would look
  like is as follows (with N being the number of yield points, m being the number of
  variables that were captured by the original code-block, and n being the number
  of parameters as defined by the type of the continuation). Note that optimizations
  can be made as to which captured variables are passed in by-copy or by-value.

The details on whether to pass them in by reference or by value are explained in Sect. 5.3.4.

```
class continuation {
  public:
    continuation(L_1 & x_1_, ..., L_m, & 1_M_)
      : 1_1(1_1_), ..., 1_m(1_m_), next_state(0) {}
    bool invoke(T_1 & p_1, ..., T_n & p_n) {
      switch (next_state) {
        case 0: return invoke_0(p_1, ..., p_n);
        case 1: return invoke_1(p_1, ..., p_n);
        ...
        case N: return invoke_N(p_1, ..., p_n);
      }
      return false;
    }
    bool invoke_0(T_1 & p_1, ..., T_n & p_n);
    bool invoke_1(T_1 & p_1, ..., T_n & p_n);
    ...
    bool invoke_N(T_1 & p_1, ..., T_n & p_n);
  private:
    integer next_state;
    L_1 & x_1; ... L_m & x_m;
};
```

- In the second step, all stack-allocated variables that are in the code-block that must remain alive across a **yield** statement, as these will no longer follow the regular scoped behavior, are turned into heap allocated ones. This includes making sure that all accesses to it are now done via pointer dereference. These are *alpha-renamed* so that they do not clash with any other variable name, either those in 1_1 ... 1_m, those in p_1 ... p_n or with each other. Graphically, this can be represented by the transformation shown in Fig. 5.5 where it shown how this only occurs when a yield statement is present somewhere inside a variable-scope, where obviously all references to X are changed to *X and an extra field is added to the continuation class to store this pointer.

- In the third step, we actually perform the continuation-passing-transformation, making sure that each **yield** statement is at the end of a block of code, and filling in each block of code in the proper invoke function, namely the code before the first yield goes in invoke_0, the code after yield i goes into invoke_i and the code after the last yield goes into invoke_N. Note that when we say *between* we do not mean syntactically, but at run-time. As such, the continuation-passing-transformation makes it so that the code that is at run-time executed in between two yield statements becomes syntactically apparent, so that we can fill in the invoke functions properly. Note that in this step, we rely on the previous step to ascertain that no longer any **yield** statements are present inside variable scopes. As such, we distinguish three cases: **yield** is in a sequence of steps; **yield** is in an if branch; or **yield** is in a while loop. Obviously, these can be nested, and therefore we use a recursive transformation. The pseudo-code is represented in Algorithm 5.1, presented in a functional language. Another syntactical construct **proceed** is introduced to determine where computation should continue in case a trace of code ends, or in other words, a function call to another method of the thunk. This is to be able to introduce extra methods so that code can be shared.

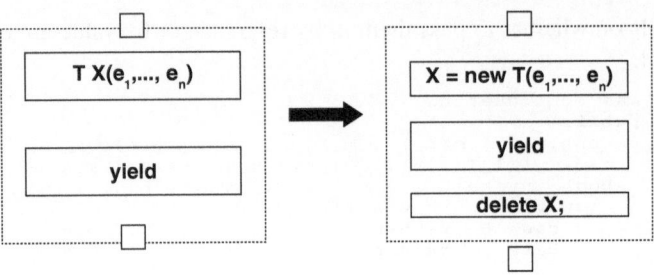

Fig. 5.5 Heap-allocation transformation

The pseudo-code in Algorithm 5.1 is presented in a functional language. To clarify presentation we have removed the curly braces in the source code. We use the type Stm to represent code, as presented before. Note that we take *act* to stand either for plain statements, or scopes that do not contain yields, as these should not be transformed. We represent a function by the datatype Function which has a name, the String parameter, and a body, the Stm parameter. A thunk, Thunk, is then simply a list of functions (plus the previously explained member variables). Since we are mixing two languages, namely SIMPLE, which we are transforming, and the pseudo-code that we use for the transformation, we delimit SIMPLE code with ≪≫ and put it in *thisfont*. As such, the SIMPLE code on the left of the -> arrows can be seen as patterns that deconstruct the code, and allow us to do a case-analysis based on the construct we are transforming. On the other hand, other code inside ≪≫ constructs new SIMPLE code that is given by the transformation. The function transform takes two parameters:

- cc : a hole that when given a SIMPLE statement, or Stm, returns the body for the current place the program is at that point as well as the rest of the program.
- *stm* : a statement

It then returns a new hole, where the next statement can be plugged into. Based on the current statement being analyzed, we can then determine where the next statement should fall. For trivial cases, such as a plain *act*, or a while or if without any yields, we can simply fill the hole we obtain with the current statement as well as the next statement, in sequence. This can be observed in the cases that return cc (≪*stm stm'* ≫). On the other hand, if we have two consecutive statements, we use the hole on the first statement, to get a new hole, and then use this new hole in transforming the second statement, to get a second hole in which we can plug in the final statement *stm'*. In the case we have a **yield** statement, we create the appropriately numbered invoke function, plug in the yield statement as the last part of the current hole, and then continue filling in this new invoke function. The next step in the method will then make sure to change these transformed yield statements in proper return statements. Note that we use the operator < + to add more functions to the current program.

The cases for **if** and **while** are slightly more complicated, as here the control-flow splits into two different directions and then merges again. To not duplicate code in the output, we therefore introduce similar join-points by creating new functions. In the case of **if** we create a new function that is uniquely named, and use as final statement for both branches a **proceed** call to this function. The statement after the **if** statement is then further transformed into this new function. Additionally, any functions that either branch of the **if** statement has produced is added to the output. In the case of the **while**, we place the join-point right at the start of the while-loop, as such the current while loop is replaced by a **proceed** to this new function. The while loop itself is then transformed into an **if** statement, with the else branch being filled with what came after the loop. The body is then transformed independently, with as final statement, a call back to the function representing this while loop.

Finally, to transform a full program, it is instantiated with the initial environment that will generate the `invoke_0` function, and is given as last statement the **finish** statement, which signifies the end of the reset-point.

- Finally, in the fourth step, we replace the **yield** statement with some assignments and a **return** statement and we add a final return at the end of the reset block. More specifically, each **yield**(e_1, ..., e_m) with i being its number, becomes:

```
p_1 = e_1; ... p_m = e_m;
next_state = i;
return true;
```

Additionally, **proceed** n becomes:

```
return n(p_1, ... p_n);
```

Finally, **finish** becomes:

```
next_state = N+1;
return false;
```

5.3.3 Demonstration

To make the above transformation more concrete, we now give an example and illustrate the different steps on a small piece of code. We take the code below, assuming that the array a is alive across this block of code. Clearly we have a production and consumption loop.

```
typedef vector<int> DataSequence;
DataSequence data;
for (int i = 0; i < n; ++i)
  if (foo(a[i]))
    data.push_back(i);

for (DataSequence::iterator j = data.begin(); j != data.end(); ++j)
  cout << *j;
```

With the rewrite rules from Sect. 5.3.1, and a proper `reset` scope choice we get the code below. We change the type definition of the vector to be a stream and then only the production code is modified. The consumption-code remains as is, although

Algorithm 5.1 Continuation Passing Transformation

```
data Function = Function String Stm
type Thunk    = [Function]
type Hole     = Stm -> (Stm, Thunk)

(<+) : : (Stm, Thunk) -> Thunk -> (Stm, Thunk)
(stm, prog) <+ procs = (stm, prog ++ procs)

newhole : : Stm -> (Stm, Thunk)
newhole stm = (stm, [])

transform : : Hole -> Stm -> Hole
transform cc stm stm' =
  case stm of
    << act >> -> cc (<< stm stm' >>)
    << a b >> ->
      let cc' = transform cc a
        in      transform cc' b stm'
    <<< yield > i (e₁,...,eₙ) >> ->
      cc (<<< yield > i (e₁,...,eₙ) >>)
          <+ [Function ("invoke" ++ show i) stm']
    <<< if > i e < then > a < else > b >> ->
      if containsYield a || containsYield b
        then
          let procname    = "procif" ++ show i
              (a', aprocs) = transform newhole a <<< proceed > procname >>
              (b', bprocs) = transform newhole b <<< proceed > procname >>
            in cc <<< if > i e < then > a' < else > b' >>
                  <+ aprocs <+ bprocs
                  <+ [Function procname stm']
        else cc (<< stm stm' >>)
    <<< while > i e < do > s >> ->
      if containsYield b
        then
          let procname    = "procwhile" ++ show i
              (b', wprocs) = transform newhole b <<< proceed > procname >>
            in cc <<< proceed > procname >>
                  <+ wprocs
                  <+ [Function procname <<< if > i e < then > b' < else > stm' >>]
        else cc (stm stm')

cps : : Stm -> Thunk
cps stm = transform (\ stm' -> (stm', [Function "invoke_0" stm']))
              stm <<< finish >>>
```

the definition of the sequence `data` is moved closer to the consumption-site, basically after the last production-site. Since we work with generator-transformations, we need an `empty` generator to start the chain. Note that the delayed code introduces a new type that we here name `DataThunk`. This will be the name of the continuation object where the code will be stored in.

```
typedef stream_vector<int> DataSequence;
DataThunk data_thunk = reset {
  for (int i = 0; i < n; ++i)
    if (foo(a[i]))
      yield(i);
}
Generator<int>                          empty;
```

```
implicit_generator<int, DataThunk> data_generator(empty, data_thunk);
DataSequence data(data_generator);

for (DataSequence::iterator j = data.begin(); j != data.end(); ++j)
    cout << *j;
```

All that remains to occur now is to hoist the `reset` code out of this context and to CPS-transform it into normal C++ code. We therefore only show the result of the definition of `DataThunk` and no longer repeat the rest of the code shown above as that remains unchanged. Since we only have one yield statement, the typing rules are quite obvious in that we only get one parameter, namely an `int`. Additionally, we only have one **yield** statement, so we should get only two numbered invoke functions, one for the start of the `reset` block and one for right after the **yield** statement. We only capture one variable in the surrounding environment, namely the array a (which we will assume for this example to be of type int a[]). Since this is an array (or pointer), it is not required to store it or pass it in as reference. In fact, this holds for all POD (plain old data types) that are constructor-less and are cheap to copy.

```
class DataThunk {
  public:
    DataThunk(int * a_)
        : a(a_), next_state(0) {}
    bool invoke(int & p_1) {
      switch (next_state) {
        case 0: return invoke_0(p_1);
        case 1: return invoke_1(p_1);
      }
      return false;
    }
    bool invoke_0(int & p_1);
    bool invoke_1(int & p_1);
  private:
    integer next_state;
    int * a;
};
```

If we now desugar the **for** loop into a while-loop (which we will skip for conciseness purposes), we see that the variable i has a scope that lives across a **yield** statement. As such it should be heap-allocated and a pointer for it should be created in the class. Here too we can introduce a minor optimization by not heap-allocating POD but just extending its lifetime to that of the Thunk itself. This saves a pointer access for each access to the variable (although the pointer access would not incur much penalty as the data will most likely be cache-local). So with this, an extra **int** i field is added to `DataThunk`. All that remains to be done now is to perform the actual CPS transform on the following remaining code:

```
reset {
  i = 0;
  while (i < n) {
    if (foo(a[i]))
      yield(i);
    ++i;
  }
}
```

If we apply the transformation presented in Sect. 5.3.2.4, we get the following methods for `DataThunk`. Note that the type of any extra introduced methods will be the same as the invoke methods. Filling all this into the results, and applying the last step of the transformations to get rid of the **finish** and **proceed** keywords, we get the following final result:

```
class DataThunk {
  public:
    DataThunk(int * a_)
      : a(a_), next_state(0) {}
    bool invoke(int & p_1) {
      switch (next_state) {
        case 0: return invoke_0(p_1);
        case 1: return invoke_1(p_1);
      }
      return false;
    }
    bool invoke_0(int & p_1) {
      i = 0;
      return procwhile0(p_1);
    }
    bool procwhile0(int & p_1) {
      if (i < n) {
        if (foo(a[i])) {
          p_1 = i;
          next_state = 1;
          return true;
        } else {
          return procif0(p_1);
        }
      } else {
        next_state = 2;
        return false;
      }
    }
    bool invoke_1(int & p_1) {
      return procif0(p_1);
    }
    book procif0(int & p_1) {
      ++i;
      return procwhile0(p_1);
    }
  private:
    integer next_state;
    int * a;
    int i;
};
typedef stream_vector<int> DataSequence;
...
Generator<int>                          empty;
DataThunk data_thunk(a);
implicit_generator<int, DataThunk> data_generator(empty, data_thunk);
DataSequence data(data_generator);

for (DataSequence: :iterator j = data.begin(); j != data.end(); ++j)
  cout << *j;
```

5.3.4 Data-Flow Constraints

Due to the fact that we are delaying certain code until a later point, we are in essence changing the low-level scheduling of the memory accesses as well as any possible side-effects. This can lead to incorrect results. Here we detail a conservative set of constraints to ensure that the transformation does not introduce bugs. As a conservative rule we state that if any code-blocks related to the production of a specific sequence violate these constraints, then we do not allow the transformation for this sequence. Note that this is a conservative set of constraints and as such future work can loosen these constraints by introducing more analyses and transformations.

First and foremost, it is clear that since the creation of a **reset** point introduces a new scope, this creation cannot cross across the existence of other scopes. More specifically, any scope is either completely inside the point, or the point is completely inside the scope, or the scope and the point do not overlap. Therefore, for the constraint rules, we only have to take into account scope-blocks, where we introduce a new one for the reset-point whose constraints we are trying to check. The constraints are due to three factors:

- Writes occurring in the code-block that is delayed by the transformation to variables that are read after.
- Reads occurring in the code-block to variables that are written between the code-block and the last consumption of the sequence to which the code-block belongs.
- Variables that are captured from the surrounding scope in a code-block not having their lifetime persist until the actual consumption of the sequence to which the code-block belongs.

5.4 Experimental Results

We will now study the effects of our transformation on two different multimedia applications. The first application is a 3D-reconstruction algorithm and the second application is a physics engine.

The first case study is a 3D reconstruction algorithm that resembles 3D perception of humans, where the relative displacement between two 2D projections (i.e., one for each eye) is used to reconstruct the third dimension. The experimental results are taken from the source code that is one of the basic building blocks in many current 3D vision algorithms. More specifically, the source code under study has been extracted from the original code of the 3D image reconstruction system (see [160] for the full code of the global 3D algorithm, which contains 1.75 million lines of high level C++) and creates the mathematical abstraction from the images or related frames that is used in subsequent phases of the global algorithm. This implementation matches corners [138] detected in two subsequent frames and was chosen due to its memory usage intensive nature.

In the core of the application, point-matches are selected by comparing neighborhoods of the two frames. These points are put into a local sequence named

Table 5.1 Comparison in terms of accesses and cycle cost between original code and IVR transformed code with 1 KB L1 for the 3D reconstruction algorithm

Code Version	L1 accesses	L2 accesses	Cycle cost	Cycle gain (%)
Original code v1	9579396	280158	23641320	0
IVR code v1	9962282	0	19924564	15.7
Original code v2	5637130	280158	15756788	0
IVR code v2	6020174	0	12040348	23.6
Original code v3	682454	280158	5847436	0
IVR code v3	1065498	0	2130996	63.6
Original code v4	0	280158	4482528	0
IVR code v4	383044	0	766088	82.9

`ImageCandidates`. For each feature point of the first image, this process is repeated and the points are selected from this sequence to be put into a more global sequence `CandidateMatches`. Once all the candidate matches have been found, a second phase of the application tries to find the best candidates. It utilizes the fact that the points in `CandidateMatches` are sorted by the point in the first image. The best candidates are filtered out and stored into another sequence `BestMatches`. After the algorithm finishes, the best matches are copied into the final output sequence named `NewMatches`. Finally, the matches in `NewMatches` are copied into the structure `OldMatches`.

Disregarding the static image input and output, since these do not belong to the core benchmark, as they serve as interfaces, we see the results as shown in Table 5.1. Here we assume a 1 KB L1 scratchpad with 2 cycles access, and a 512 KB L2 memory with 16 cycles access. In both cases, the data was binpacked onto the smallest possible scratchpad for optimal performance numbers. It should be noted that the vast majority of the accesses in both the old and the new algorithm were due to three registered-sized integers, namely the loop-counter and a 2-element array that is filled on each loop iteration. Obviously, any compiler would map this directly to a register. The second version compares when these accesses are not counted. A third variable with a very high access count is present that did not change due to the transformation. As that variable is statically allocated, the DTSE transformation method [110] should be applied to reduce its impact. Here we assume that this approach has been sufficiently applied to achieve this so the third version compares the cycle cost before and after the transformation without accounting for this static variable. This gives an optimistic (but not overly optimistic) view of the achieved gains. Finally, the fourth version discounts access to any of the register-sized variables, since these will all be assigned to the foreground memory where both the cycle and the energy overhead is negligible. Table 5.2 shows the same comparison, but in this case the scratchpad is 4 KB. Clearly, as the scratchpad size increases, the number of accesses of the original version shift towards L1. Therefore the cycle gains through IVR are less pronounced. This, however, comes at the cost of more area for the larger L1 and a higher energy profile.

Table 5.2 Comparison in terms of accesses and cycle cost between original code and IVR transformed code with 4 KB L1 for the 3D reconstruction algorithm

Code Version	L1 accesses	L2 accesses	Cycle cost	Cycle gain (%)
Original code v1	9706336	153060	21861632	0
IVR code v1	9962282	0	19924564	8.8
Original code v2	5764228	153060	13977416	0
IVR code v2	6020174	0	12040348	13.9
Original code v3	809552	153060	4068064	0
IVR code v3	1065498	0	2130996	47.6
Original code v4	127098	153060	2703156	0
IVR code v4	383044	0	766088	71.7

The second case study is a physics engine that simulates elastic and deformable bodies [88]. The complexity in modeling elastic and deformable bodies over more classical physics engines that consider objects as completely static, is that all the different edges must be modeled independently, thus resulting in a high amount of data transfers. To be able to get deformable models, the different edges of a 3D objects are represented as strings with a certain constant, length and dampening. Whenever collisions occur between two objects, these result in impulses applied to the specific vertices where the collisions occur, and these forces then propagate through the spring model representing the objects. While a varied set of techniques deal hierarchically with collision among objects, once two objects are found to be colliding, the vertices involved in the collision need to be processed.

In the core logic of this simulator, for every pair of objects, the simulator checks whether these collide. And if they do, it gets a complete listing of all the vertices of each object that were affected. These are then adjusted by giving them an impulse in velocity. This results in a constant creation of vectors of vertices that are then consumed once. As such, these vectors of vertices, created and consumed very frequently, are a prime target for removal.

The actual generation of these lists of vertices happens in two phases. In a lightweight phase, it is first detected if any vertices and edges are directly affected by the collision, and thus whether two objects actually collide. The pseudo-code for this is shown in Algorithm 5.2. First `CollisionDetected` finds all the points directly involved in the collision. Then, `DFS` appends to the list of points based on the edges that were affected. Since `CollisionDetected` is quite computationally expensive, but the result is required for the if condition, it is not optimal to delay the generation of the points from this code, as this would lead to re-execution. This is an example where memoization is efficient. This nicely illustrates the trade-off between (re)computation and memoization which is important to decide prior to the decision for applying the earlier mentioned transformations (Sect. 5.3.2.4) in the IVR stage.

The IVR method then keeps the base points array generated by the function `CollisionDetected`, however all the extra points generated by `DFS` are only generated on demand, resulting in fewer memory accesses.

It is realistic to assume that register accesses do not involve any cycle overhead compared to the baseline. That is namely motivated by the presence of multiple FUs/slots on the processor [36]. So the few extra scalar operations introduced by the transformation can be mapped to the empty schedule slots, as no blocking dependencies will be present. And the register access itself is also not requiring a separate cycle. In that case, the gain in memory memory accesses on these intermediate data structures is 96.45 %. Basically, that is due to the fact that the entire vector of points is produced once and consumed once, and the majority of the produced points result from the `DFS` method.

Algorithm 5.2 Core logic of physics engine

```
for objectA in all objects:
  for objectB in all objects:
    vector<CollisionEdge> collisionEdgesA, collisionEdgesB;
    vector<GLPoint> collisionPointA, collisionPointB;
    if (objectA->CollisionDetected(
        objectB, collisionEdgesA, collisionPointB))
    {
      objectB->CollisionDetected(
        objectA, collisionEdgesB, collisionPointA);
      objectA->DFS(collisionEdgesA, collisionPointA);
      objectB->DFS(collisionEdgesB, collisionPointB);
      // Consumption of collisionPointA and collisionPointB
```

5.5 Conclusions

In this chapter we have presented a method that transforms sequence data types into stream-generators. Through the use of this transformation, it is possible to reduce the energy consumption of an application by lowering the number of memory accesses to large structures. Instead, they are replaced by procedures that generate this data on-demand. Therefore, the memory footprint is reduced, and the memory accesses are to smaller structures that can be placed in smaller memories, and thus less costly ones, in terms of energy consumption.

However, for other types of dynamic applications, the IVR can be extended to deal with deterministic access patterns that are not only based on forward consumption and the use of the operators mentioned in the Sect. 5.2.1. If index-based operators are used, then instead traditional approaches such as the polyhedral model are best used, and if sequences are partially used in a linear fashion, and partially through random-access, then perhaps some sort of hybrid approach is required.

Chapter 6
Dynamic Memory Management Optimization for Multimedia Applications

As already introduced in the first two chapters of this book, due to increasing complexity and drastic rise in memory requirements, new system-level memory management methodologies for multimedia applications need to be developed. In the past, most applications that were ported to embedded platforms stayed mainly in the classic domain of signal processing and actively avoided algorithms that employ data de/allocated dynamically at run-time, also called Dynamic Memory (DM from now on). Recently, the multimedia and wireless network applications to be ported to embedded systems have experienced a very fast growth in their variety, complexity and functionality. These applications (e.g., MPEG4 or new network protocols) depend, with few exceptions, on DM for their operations due to the inherent unpredictability of the input data. Designing the final embedded systems for the (static) worst case memory footprint[1] of these applications would lead to a too high overhead for them. Even if average values of possible memory footprint estimations are used, these static solutions will result in higher memory footprint figures (i.e., more approximately 25 %) than DM solutions [98]. Furthermore, these intermediate static solutions do not work in extreme cases of input data, while DM solutions can do it since they can scale the required memory footprint. Thus, DM management mechanisms must be used in these designs.

Many general DM management policies, and implementations of them, are nowadays available to provide relatively good performance and low fragmentation for general-purpose systems [171]. However, for embedded systems, such managers must be implemented inside their constrained Operating System (OS) and thus have to take into account the limited resources available to minimize memory footprint among other factors. Thus, recent embedded OSes (e.g., [128]) use custom DM managers according to the underlying memory hierarchy and the kind of applications that will run on them.

Usually custom DM managers are designed to improve performance [27, 171], but they can also be used to heavily optimize memory footprint compared to general-purpose DM managers, which is very relevant for final energy and performance

---

[1] Accumulated size of all the data allocated in memory and counted in bits.

© Springer International Publishing Switzerland 2015
D. Atienza Alonso et al., *Dynamic Memory Management for Embedded Systems*,
DOI 10.1007/978-3-319-10572-7_6

in embedded systems as well since many concurrent dynamic applications have to share the limited on-chip memory available. For instance, in 3D vision algorithms [138], a suitably designed custom DM manager can improve memory footprint by 45 % approximately over conventional general-purpose DM managers [98]. However, when custom DM managers are used, their designs have to be manually optimized by the developer, typically considering only a limited number of design and implementation alternatives, which are defined based on his experience and inspiration. This limited exploration is mainly restricted due to the lack of systematic methodologies to consistently explore the DM management design space. As a result, designers must define, construct and evaluate new custom implementations of DM strategies manually, which has been proved to be programming intensive (and very time-consuming). Even if the embedded OS offers considerable support for standardized languages, such as C or C++, the developer is still faced with defining the structure of the DM manager and how to profile it on a case per case basis.

In this chapter, we present a methodology that allows developers to design custom DM management mechanisms with the reduced memory footprint required for these dynamic multimedia and wireless network applications. First of all, this methodology delimits the relevant design space of DM management decisions for a minimal memory footprint in dynamic embedded applications. After that, we have studied the relative influence of each decision of the design space for memory footprint and defined a suitable order to traverse this design space according to the DM behavior of these dynamic applications. As a result, the main contributions of our methodology are two-fold: (i) the definition of a consistent orthogonalization of the design space of DM management for embedded systems and (ii) the definition of a suitable ordering of design decisions for dynamic multimedia and wireless network applications (and any other type of embedded applications that possesses the same dynamic de/allocation characteristics) to help designers to create very customized DM managers according to the specific dynamic behavior of each application.

This chapter is organized as follows. In Sect. 6.1, we describe relevant related work in the context of memory management for multimedia applications. In Sect. 6.2 we present the relevant DM management design space of decisions for a reduced memory footprint in dynamic multimedia applications. In Sect. 6.3 we define the order to traverse this design space to minimize the memory footprint of the application under analysis. In Sect. 6.4, we outline the global design flow proposed in our methodology to minimize the memory footprint in dynamic embedded applications. In Sect. 6.5, we introduce our case studies and present in detail the experimental results obtained. Finally, in Sect. 6.6 we draw our conclusions.

6.1 Related Work

The basic principles of an efficient DM management in a general-context are already well established. Much literature is available about general-purpose DM management software implementations and policies [171]. Trying to reuse this extensive available literature, embedded systems where the range of applications to be executed is very wide (e.g., consumer devices) tend to use variations of well-known state-of-the-art general-purpose DM managers. For example, Linux-based systems use as their basis the Lea DM manager [27, 171] and Windows-based systems (both mobile and desktop) include the ideas of the Kingsley DM manager [43, 44, 171]. Finally, some real-time OSes for embedded systems (e.g. [128]) support DM de/allocation via custom DM managers based on simple region allocators [66] with a reasonable level of performance for the specific platform features. All these approaches propose optimizations considering general-purpose systems, where the range of applications are very broad and unpredictable at design time. In contrast, our approach takes advantage of the special DM behavior of multimedia and wireless network applications to create highly customized and efficient DM managers for embedded systems.

Later, research on custom DM managers that take application-specific behavior into account to improve performance appeared [27, 169, 171]. Also, for improving speed in highly constrained embedded systems, [127] proposes to partition the DM into fixed blocks and place them in a single linked list with a simple (but fast) fit strategy, e.g., first fit or next fit [171]. In addition, some partially configurable DM manager libraries are available to provide low memory footprint overhead and high level of performance for typical behaviors of a certain application (e.g., Obstacks [171] is a custom DM manager optimized for a stack-like allocation/de-allocation behavior). Similarly, [169] proposes a DM manager that allows to define multiple regions in memory with several user-defined functions for memory de/allocation. Additionally, since the incursion in embedded systems design of object-oriented languages with support for automatic recycling of dead-objects in DM (usually called garbage collection), such as Java, work has been done to propose several automatic garbage collection algorithms with relatively limited overhead in performance, which can be used in real-time systems [28, 51]. In this context, also hardware extensions have been proposed to perform garbage collection more efficiently [154]. The main difference of these approaches compared to ours is that they mainly aim at performance optimizations and propose ad-hoc solutions without defining a complete design space and exploration order for dynamic embedded systems as we propose in this chapter.

In addition, research has been performed to provide efficient hardware support for DM management. [38] presents an Object Management Extension (i.e., OMX) unit to handle the de/allocation of memory blocks completely in hardware using an algorithm which is a variation of the classic binary buddy system. [146] proposes a hardware module called SoCDMMU (i.e., System-On-a-Chip Dynamic Memory Management Unit) that tackles the global on-chip memory de/allocation to achieve a deterministic way to divide the memory among the processing elements of SoC designs. However, the OS still performs the management of memory allocated to a particular on-chip processor. All these proposals are very relevant for embedded systems where the hardware can still be changed, while our work is thought for fixed embedded design architectures where customization can only be done at the OS or software level.

Finally, a lot of research has been performed in optimization techniques to improve memory footprint, power consumption, performance and other relevant factors in static data for embedded systems (see surveys in [25, 132]). All these techniques are complementary to our work and are usable for the static data that usually are also present in the dynamic applications we consider.

6.2 Relevant Design Space for Dynamic Memory Management in Dynamic Multimedia and Wireless Network Applications

In the software community much literature is available about possible design choices for DM management mechanisms [171], but none of the earlier work provides a complete design space useful for a systematic exploration in multimedia and wireless network applications for embedded systems. Hence, in Sect. 6.2.1 we first detail the set of relevant decisions in the design space of DM management for a reduced memory footprint in dynamic multimedia and wireless network applications. Then, in Sect. 6.2.2 we briefly summarize the inter-dependencies observed within this design space, which partially allow us to order this vast design space of decisions. Finally, in Sect. 6.2.3 we explain how to create global DM managers for dynamic multimedia and wireless network applications.

6.2.1 Dynamic Memory Management Design Space for Reduced Memory Footprint

Conventional DM management basically consists of two separate tasks, namely allocation and de-allocation [171]. In particular, allocation is the mechanism that searches for a block big enough to satisfy the request of a given application and de-allocation is the mechanism that returns this block to the available memory of the system in order to be reused later by another request. In real applications, the blocks can be requested and returned in any order, thus creating "holes" among used blocks. These holes are known as *memory fragmentation*. On the one hand, internal fragmentation occurs when a bigger block than the one needed is chosen to satisfy a request. On the other hand, if free memory to satisfy a memory request is available, but not contiguous (thus it cannot be used for that request), it is called external fragmentation. Therefore, on top of memory allocation and de-allocation, the DM manager has to take care of fragmentation issues as well. This is done by splitting and coalescing free blocks to keep memory fragmentation as small as possible. Finally, to support all these mechanisms, additional data structures should be built to keep track of all the free and used blocks, and the defragmentation mechanisms. As a result, to create an efficient DM manager, we have to systematically classify the design decisions that can be taken to handle all the possible combinations of the factors that affect the DM subsystem (e.g., fragmentation, overhead of the additional data structures, etc.).

We have classified all the important design options that constitute the design space of DM management in different orthogonal decision trees. Orthogonal means that any decision in any tree can be combined with any decision in another tree, and the result should be a potentially valid combination, thus covering the whole possible design space. Then, the relevance of a certain solution in each concrete system depends on its design constraints, which implies that some solutions may not meet all timing and cost constraints for that concrete system. Furthermore, the decisions in different orthogonal trees can be sequentially ordered in such a way that the trees can be traversed without iterations, as long as the appropriate constraints are propagated from one decision level to all subsequent levels. Basically, when one decision has been taken in every tree, one custom DM manager is defined (in our notation, atomic DM manager) for a specific DM behavior pattern (usually, one of the DM behavior phases of the application). In this way we can recreate any available general purpose DM manager [171] or create our own highly-specialized DM managers.

The trees have been grouped in categories according to the different main parts that can be distinguished in DM management [171]. An overview of the relevant classes of this design space for a reduced memory footprint is shown in Fig. 6.1. This approach allows to reduce the complexity of the global design of DM managers in smaller subproblems that can be decided locally, making feasible the definition of a convenient order to traverse it.

In the following we describe the five main categories and decision trees shown in Fig. 6.1. For each of them we focus on the decision trees that are important for the creation of DM managers with a reduced memory footprint.

- A. *Creating block structures*, which handles the way block data structures are created and later used by the DM managers to satisfy the memory requests. More specifically, the *Block structure* tree specifies the different blocks of the system and their internal control structures. In this tree, we include all possible combinations of Dynamic Data Types (from now on called DDTs) required to represent and construct any dynamic data representation [52, 98] used in a DM manager. Secondly, the *Block sizes* tree refers to the different sizes of basic blocks available for DM management, which may be fixed or not. Thirdly, the *Block tags* and the *Block recorded info* trees specify the extra fields needed inside the block to store information used by the DM manager. Finally, the *Flexible block size manager* tree decides if the splitting and coalescing mechanisms are activated or extra memory is requested from the system. This depends on the availability of the size of the memory block requested.
- B. *Pool division based on criterion*, which deals with the number of pools (or memory regions) present in the DM manager and the reasons why they are created. The *Size* tree means that pools can exist either containing internally blocks of several sizes or they can be divided so that one pool exists per different block size. In addition, the *Pool structure* tree specifies the global control structure for the different pools of the system. In this tree we include all possible combinations of DDTs required to represent and construct any dynamic data representation of memory pools [52, 98].

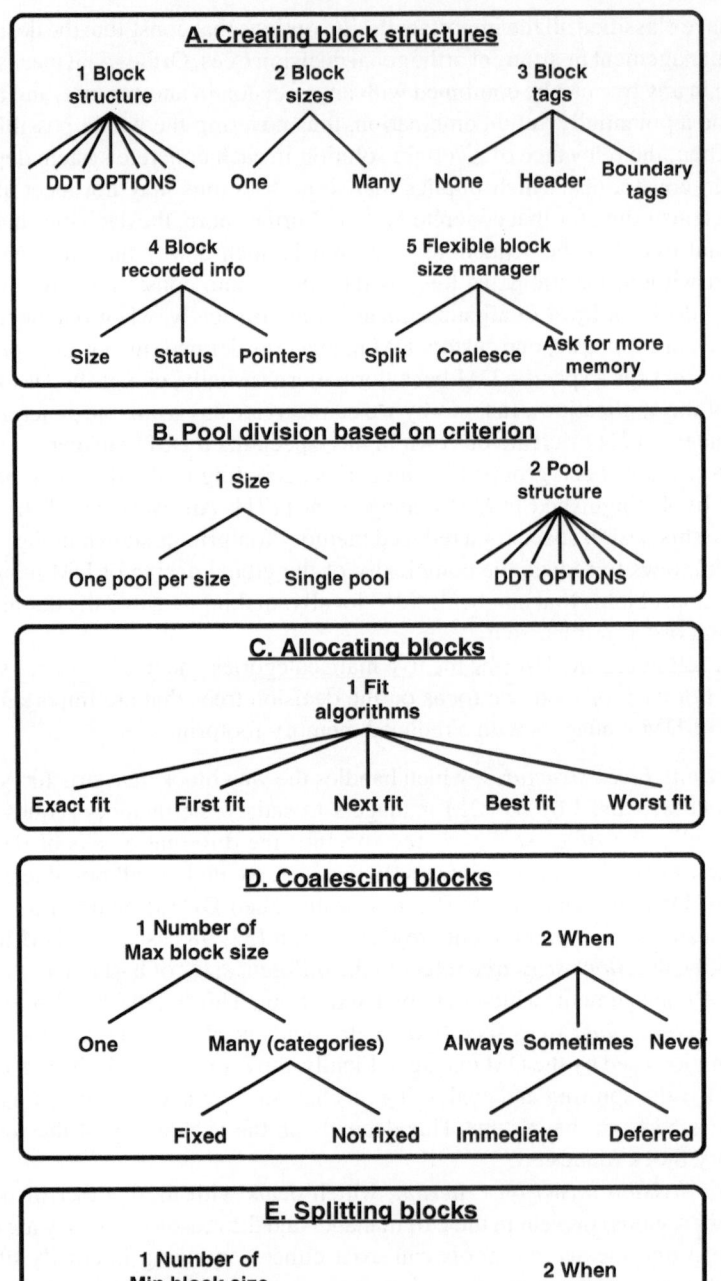

Fig. 6.1 Overview of the DM management design space of orthogonal decisions for reduced memory footprint

- *C. Allocating blocks*, which deals with the actual actions required in DM management to satisfy the memory requests and couple them with a free memory block. Here we include all the important options available in order to choose a block from a list of free blocks [171]. Note that a De-allocating blocks category with the same trees as this category could be created, but we do not include it in Fig. 6.1 to avoid adding complexity unnecessarily to the DM management design space. The fact is that the Allocating blocks category possesses more influence for memory footprint than the additional De-allocating blocks category. Moreover, these two categories are so tightly linked together regarding memory footprint of the final solution that the decisions taken in one must be followed in the other one. Thus, the De-allocating blocks category is completely determined after selecting the options of this Allocating block category.
- *D. Coalescing blocks*, which is related to the actions executed by the DM managers to ensure a low percentage of external memory fragmentation, namely merging two smaller blocks into a larger one. First, the *Number of max block size* tree defines the new block sizes that are allowed after coalescing two different adjacent blocks. The *When* tree defines how often coalescing should be performed.
- *E. Splitting blocks*, which refers to the actions executed by the DM managers to ensure a low percentage of internal memory fragmentation, namely splitting one larger block into two smaller ones. First, the *Number of min block size* tree defines the new block sizes that are allowed after splitting a block into smaller ones. And the *When* tree defines how often splitting should be performed (these trees are not presented in full detail in Fig. 6.1, because the options are the same as in the two trees of the Coalescing category).

6.2.2 Interdependencies Between the Orthogonal Trees

After this definition of the decision categories and trees, in this section we identify their possible interdependencies. They impose a partial order in the characterization of the DM managers. The decision trees are orthogonal, but not independent. Therefore, the selection of certain leaves in some trees heavily affects the coherent decisions in the others (i.e., interdependencies) when a certain DM manager is designed. The whole set of interdependencies for the design space is shown in Fig. 6.2. These interdependencies can be classified in two main groups. First, the interdependencies caused by certain leaves, trees or categories, which disable the use of other trees or categories (drawn as full arrows in Fig. 6.2). Second, the interdependencies affecting other trees or categories due to their linked purposes (shown as dashed arrows in Fig. 6.2). The arrows indicate that the starting side affects the receiving one and must be decided first.

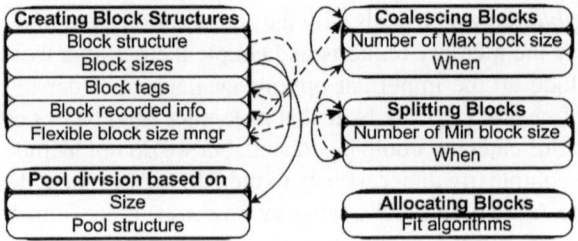

Fig. 6.2 Interdependencies between the orthogonal trees in the design space

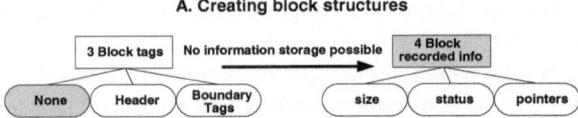

Fig. 6.3 Example of interdependency between two orthogonal trees of the DM management design space

6.2.2.1 Leaves or Trees that Obstruct the Use of Others in the Design Space

These interdependencies appear due to the existence of opposite leaves and trees in the design space. They are depicted in Fig. 6.2 and are represented by *full* arcs:

- First, inside the Creating block structures category, if the *none* leaf from the *Block tags* tree is selected, then the *Block recorded info* tree cannot be used: there would be no memory space to store the recorded info inside the block. This interdependency is graphically shown as example in Fig. 6.3.
- Secondly, the *one* leaf from the Block sizes tree excludes the use of the *Size* tree in the Pool division based on criterion category and of the *Flexible block size manager* tree in the Creating block structures category. This occurs because the *one block size* leaf does not allow to define any new block sizes.

6.2.2.2 Leaves or Trees that Limit the Use of Others in the Design Space

These interdependencies exist since the leaves have to be combined to create consistent whole DM schemes. For example, the coalescing and splitting mechanisms are quite related and the decisions in one category have to find equivalent ones in the other one. These interdependencies are represented with *dashed* arcs in Fig. 6.2.

- First, the *Flexible block size manager* tree heavily influences all the trees inside the Coalescing Blocks and the Splitting Blocks categories. Thus, according to the selected leaf for a certain atomic DM manager (i.e., the *split* or *coalesce* leaf), the DM manager has to select some leaves of the trees involved in those decisions or not. For example, if the *split* leaf is chosen, the DM manager will not use the coalescing category and the functions of their corresponding trees inside that atomic the DM manager. However, it can be used in another atomic DM manager in a different DM allocation phase, thus the final global manager will contain both (see Sect. 6.2.3 for more details).

However, the main cost of the selection done in the *Flexible block size manager* tree is characterized by the cost of the Coalescing Blocks and Splitting Blocks categories. This means that a coarse grain estimator of their cost influence must be available to take the decision in the aforementioned *Flexible block size manager* tree. In fact, this estimator is necessary whether the derived decisions in other categories have an influence in the final cost not smaller than the influence of the tree that decides to use them or not [34]. This does not happen at the tree ordering and thus no estimator is required.

- Secondly, the decision taken in the *Pool structure* tree significantly affects the whole Pool division based on criterion category. This happens because some data structures limit or do not allow the pool to be divided in the complex ways that the criteria of this category suggest.
- Thirdly, the *Block structures* tree inside the Creating block structures category strongly influences the decision in the *Block tags* tree of the same category because certain data structures require extra fields for their maintenance. For example, single linked lists require a *next* field and a list where several block sizes are allowed has to include a *header* field with the size of each free block inside. Otherwise the cost to traverse the list and find a suitable block for the requested allocation size is excessive [52].
- Finally, the respective *When* trees from the Splitting and Coalescing Blocks categories are linked together with a double arrow in Fig. 6.3 because they are very tightly related to each other and a different decision in each of these two trees does not seems to provide any kind of benefit to the final solution. On the contrary, according to our study and experimental results it usually increases the cost in memory footprint of the final DM manager solution. However, this double arrow is needed because it is not possible to decide which category has more influence in the final solution without studying the specific factors of influence for a certain metric to optimize (e.g., memory footprint, power consumption, performance, etc.). Thus, the decision about which category should be decided first has to be analyzed for each particular cost function or required metric to optimize in the system.

6.2.3 Construction of Global Dynamic Memory Managers

Modern multimedia and wireless network applications include different DM behavior patterns, which are linked to their logical phases (see Sect. 6.5 for real-life examples). Consequently, our methodology must be applied to each of these different phases separately in order to create an atomic custom DM manager for each of them. Then, the global DM manager of the application is the inclusion of all these atomic DM managers in one. To this end, we have developed a C++ library based on abstract classes or templates that covers all the possible decisions in our DM design space and enables the construction of the final global custom DM manager implementation in a simple way via composition of C++ layers [12].

6.3 Order for Reduced Dynamic Memory Footprint in Dynamic Multimedia and Wireless Network Applications

Once the whole design space for DM management in dynamic embedded systems has been defined and categorized in the previous section, the order for different types of applications can be defined according to their DM behavior and the cost function/s to be optimized. In this case, the DM subsystem is optimized to achieve solutions with a reduced DM footprint. Therefore, first, in Sect. 6.3.1 we summarize the factors of influence for DM footprint. Then, in Sect. 6.3.2 we briefly describe the features that allow us to group different dynamic applications and focus on the common (and particular) features of multimedia and wireless network applications, which enable to cluster these applications and define a common exploration order of the design space. Finally, in Sect. 6.3.3 we present the suitable exploration order for these multimedia and wireless network applications to attain reduced memory footprint DM management solutions.

6.3.1 Factors of Influence for Dynamic Memory Footprint Exploration

The main factors that affect memory size are two: the *Organization overhead* and the *Fragmentation memory waste.*

1. The *Organization overhead* is the overhead produced by the assisting fields and data structures, which accompany each block and pool respectively. This organisation is essential to allocate, de-allocate and use the memory blocks inside the pools, and depends on the following parts:

   - The fields (e.g., headers, footers, etc.) inside the memory blocks are used to store data regarding the specific block and are usually a few bytes long. The use of these fields is controlled by category A (Creating block structures) in the design space.
   - The assisting data structures provide the infrastructure to organize the pool and to characterize its behavior. They can be used to prevent fragmentation by forcing the blocks to reserve memory according to their size without having to split and coalesce unnecessarily. The use of these data structures is controlled by category B (Pool division based on criterion). Note that the assisting data

structures themselves help to prevent fragmentation, but implicitly produce some overhead. The overhead they produce can be comparable to the fragmentation produced by small data structures. Nonetheless this negative factor is overcome by their ability to prevent fragmentation problems, which is a more relevant negative factor for memory footprint [171]. The same effect on fragmentation prevention is also present in tree C1. This happens because depending on the fit algorithm chosen, it is possible to reduce the internal fragmentation of the system. For example, if a 12-byte block is allocated using a next fit algorithm and the next block inside the pool is 100 bytes, 88 bytes are lost in internal fragmentation while a best fit algorithm will probably never use that block for the request. Therefore, in terms of preventing fragmentation, category C is equally important to B.

2. The *Fragmentation memory waste* is caused by the internal and external fragmentation, discussed earlier in this chapter, which depends on the following:

- Internal fragmentation is mostly remedied by category E (Splitting blocks). This mostly affects to small data structures. E.g., if you have only 100-byte blocks inside your pools and you want to allocate 20-byte blocks, it would be wise to split each 100-byte block inside your pool to 5 blocks of 20 bytes to avoid too much internal fragmentation.
- External fragmentation is mostly remedied by category D (Coalescing blocks). It mostly affects to big data requests. For example, if a 50 KB block needs to be allocated, but there are only 500 byte blocks inside the pools, it would be necessary to coalesce 100 blocks to provide the requested amount of memory.

Note the distinction between categories D and E, which try to deal with fragmentation, as opposed to category B and C that try to prevent it.

6.3.2 Analysis of de/allocation Characteristics of Dynamic Embedded Multimedia and Wireless Network Applications

Dynamic multimedia and wireless network embedded applications involve several de/allocation phases (or patterns) for their data structures, which usually represent different phases in the logic of the application itself. We have classified these different DM allocation patterns [171] in three orthogonal components, namely, *Predominant allocation block size, Main pattern* and *Proportion of intensive memory de/allocation phases.*

As we have empirically observed, using this previous classification based on components, dynamic multimedia and wireless network applications share the main features regarding DM footprint. First, the predominant allocation block sizes are few (as our case studies indicate, 6 or 7 can account for 70–80 % of the total allocation requests), but these sizes display a large variation since some of them can be just a few bytes whereas others can be several Kbytes. Therefore, DM managers suitable for only large or very small sizes are not appropriate, and combinations of solutions for both types of allocations are required in these applications. Second, the *main pattern* component, which defines the dominant pattern of de/allocations, e.g., ramp, peaks, plateaus, (see [171] for more details), indicates that very frequently dynamic multimedia and wireless network applications possess very active phases where few data structures are dynamically allocated in a very fast and variable way (i.e., peaks) while other data structures grow continuously (i.e., ramps) and remain stable at a certain point for a long time before they are freed. According to our observations, these phases of creation/destruction of sets of data structures with certain allocation sizes are mostly influenced by the logical structure of phases defined by the designers in the application code (e.g., rendering phases) or special events (e.g., arrival of packets in wireless network applications). Third, the proportion of intensive memory de/allocation phases defines how often the structure of the DM manager has to be changed to fit the new de/allocation sizes and pattern of the dynamic application. In this case we have also observed very similar features in both dynamic multimedia and wireless network applications because the run-time behavior of the system tends to follow a predefined order (i.e., defined by the designer) about how to handle the different phases and events. For example, in 3D games the sequence to service new events, as the movement of the camera by the players (e.g., update of the visible objects, rendering of the new background, etc.) is always fixed depending on the type of object. Similarly, the way to handle the arrival of new packets in wireless network applications is also fixed. Thus, all these previous features allow us to cluster these two initially different fields of applications considering their shared DM characteristics and define a common order to traverse the design space.

6.3.3 Order of the Trees for Reduced Memory Footprint in Dynamic Multimedia and Wireless Network Applications

In this subsection we discuss the global order inside the orthogonal categories of our DM management design space according to the aforementioned factors of influence for a reduced memory footprint for dynamic multimedia and wireless network applications. We have defined this global order after extensive testing (i.e., more than 6 months of experiments with different orders and implementations of DM managers) using a representative set of ten real-life dynamic embedded multimedia and wireless network applications with different code sizes (i.e., from 1000 lines to 700 K lines of C++ code), including: scalable 3D rendering such as [104] or MPEG 4 Visual Texture Coder (VTC) [124], 3D image reconstruction algorithms [138], 3D games [71, 122] and buffering, scheduling and routing network applications [115].

Experience suggests that most of the times fragmentation cannot be avoided only with a convenient pool organization [171]. Our experiments show that this holds especially true for embedded dynamic multimedia and wireless network applications. Therefore, categories D and E are placed in the order before categories C and B. The final order is as follows: the tree A2 is placed first to determine if one or more block sizes are used and A5 is placed second to decide the global structure of the blocks.

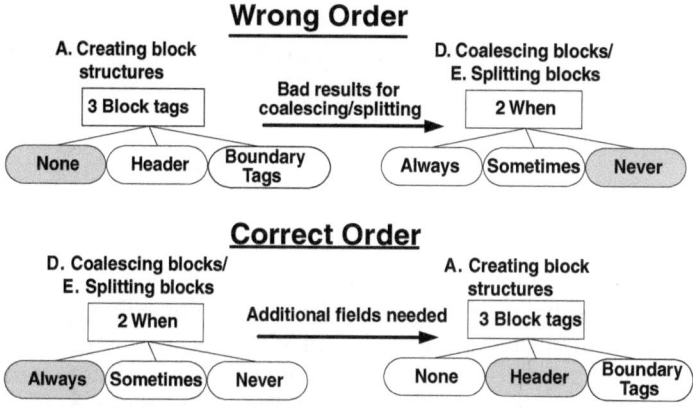

Fig. 6.4 Example of correct order between two orthogonal trees

Next, categories that deal with fragmentation, i.e., categories D and E, are decided because, as we have mentioned, they are more important than categories that try to prevent fragmentation (i.e., C and B). Then, the rest of the organization overhead must be decided for these block requests. Thus, the rest of the trees in category A (i.e., A1, A3 and A4) are decided. As a result, taking into account the interdependencies, the final global order is the following: A2->A5->E2->D2->E1->D1->B4->B1->C1->A1->A3->A4.

If the order we have just proposed is not followed, unnecessary constraints are propagated to the next decision trees. Hence, the most suitable decisions cannot be taken in the remaining orthogonal trees. An example of this is shown in Fig. 6.4. Suppose that the order was A3 and then E2 and D2. When deciding the correct leaf for A3, the obvious choice to save memory footprint would be to choose the *None* leaf, which indicates that no header fields of any type should be used. This seems reasonable at first sight because the header fields would require a fixed amount of additional memory for each block that is going to be allocated. After the decision about the *None* leaf in tree A3, the leaves to use for the trees E2 and D2 are decided. Now, we are obliged to choose the *Never* leaf because after propagating the constraints of A3, blocks cannot be properly split or coalesced without storing information about their size. Hence, the final DM Manager uses less memory per block, but cannot deal with internal or external fragmentation by splitting or coalescing blocks. This solution could be seen as an appropriate decision for an application where the predominant block size is fixed and thus no serious effects exist due to internal or external fragmentation. However, if the application includes a variable amount of allocation sizes (typical behavior in many dynamic multimedia and wireless network applications), the fragmentation problem (internal and external) is going to consume more memory than the extra header fields needed for coalescing and splitting since the freed blocks will not be able to be reused. Therefore, it is necessary to decide the E2 and D2 trees first, and then propagate the resulting constraints to tree A3.

Table 6.1 Example of two DM managers with different order in the DM design space

DM managers	Memory footprints (KB)	Execution time (s)
(1) A3-D2/E2	2.392×10^2	7.005
(2) D2/E2-A3	4.682×10^1	11.687
Comparison 2–1	19.56 %	166.83 %

To demonstrate the correctness of this explanation, in the following paragraphs the experimental results of memory footprint for a multimedia application with a varying predominant block size that uses each of the possible orders are shown, i.e., a 3D algorithm with scalable meshes [104] that is later described more in detail in our case studies and experimental results (Sect. 6.5). The results shown are average values after 10 runs for each experiment and groups of 10 bursts. First of all, the application is divided in several bursts of 1,000 (de)allocations with a limited amount of blocks (10 different sizes that range from 20 bytes to 500 bytes). This is a typical way in which a multimedia application with several phases can run. In network applications the range of sizes can vary even more due to the uncertainty in the workload and input packets. We have implemented two different versions of the same DM manager, which include a single pool where both the free and used blocks are doubly linked inside one list. The only difference between these two DM managers is the aforementioned explained order between A3 and D2/E2.

In the first solution, A3 is decided and then D2 and E2. Hence, no coalescing and splitting services are provided in the DM manager, but also less overhead in the blocks is required. In the second solution D2 and E2 are decided first and then A3 is determined. Thus, this DM manager includes both services (i.e., coalescing and splitting), but includes an overhead of 7 bytes per block in the list as additional header for maintenance: 2 bytes for the size, each link 2 bytes (2 required) and 1 byte (i.e., a bool field) to decide if it is used. The results obtained in both cases are shown in Table 6.1. It shows clearly that the overhead of the additional headers is less significant than the overhead due to fragmentation, where 5 times less memory is required (see line Comparison 2-1 in Table 6.1). Therefore, as we propose in our global exploration order for reduced memory footprint (see Sect. 6.3.3 for more details), the second DM manager produces better results. In this second manager, first the D2 and E2 trees are decided and then their constraints are used to select the A3 tree leaf. Also note the difference in execution time: the first DM manager has a very low total execution time because it does not provide coalescing or splitting services, while the second one has an additional overhead in execution time of 66.83 %. This shows that the performance requirement of the system under development must be taken into account when the choices in the design space are decided. Hence, if the system designer requires a certain level of performance, which is not achieved by the extreme reduced memory footprint solution shown in number 2, another solution closer to solution number 1 (but with a more balanced trade-off between memory footprint and performance) can be designed. As it has been mentioned before, we have performed a large amount of similar experiments for the rest decision orders and derived the proposed one. The main conclusion from our studies is that with the proposed global order the exploration methodology is not unnecessarily limited and only the necessary constraints are forwarded to the next decision trees, which is not possible with other orders for these types of dynamic applications.

6.4 Overview of the Global Flow in the Dynamic Memory Management Methodology

The main objective of the proposed DM management design methodology is to provide developers with a complete design flow of custom DM managers for dynamic multimedia and wireless network systems. This design flow is divided into four main phases, as indicated in Fig. 6.5.

The first phase of the methodology obtains detailed information about the DM sizes requested and the time when each dynamic de/allocation is performed, as depicted in the first oval shape of Fig. 6.5. This phase is based on profiling the use

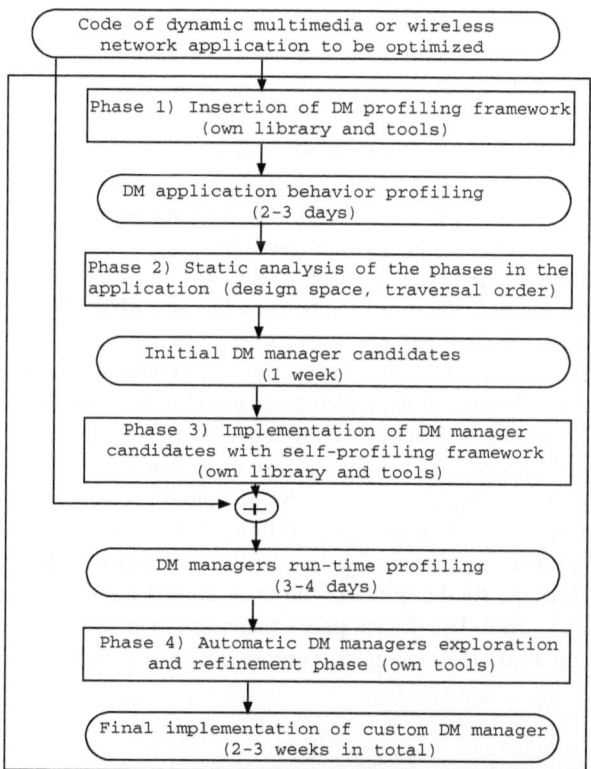

Fig. 6.5 Global overview of the proposed DM management design methodology flow

of DM in the considered application because, according to our experience, it is the only realistic option for today's highly-dynamic multimedia and wireless network applications [12, 98]. Profiling is carried out as follows. First, we insert in the code of the application our own high-level profiling framework, which is based on a C++ library of profiling objects and additional graphical parsing tools [98]. Then, we automatically profile a representative set of input cases of the application, in general between 10 and 15 different inputs, including the extreme cases of least and maximum memory footprint. This small set may be provided by the designers of the application since they use it to create their dynamic data structures to store and retrieve the data in the application (e.g., dynamic arrays, doubly linked lists, binary trees, etc.) [52]. As a result, this first phase takes between 2–3 days in our methodology for real-life applications thanks to our tools with very limited user interaction.

Then, in the second phase of the methodology, represented as the second rectangular shape of Fig. 6.5, the DM behavior of the application is characterized. The de/allocation phases are identified and linked to the logical phases of the application (e.g., rasterization, clipping, illumination, etc. in 3D rendering [172]) using our tools to analyze the profiling information obtained in the previous phase. These tools identify in different graphs and tables the main variations in memory footprint and the dynamically de/allocated sizes. Then, using the design space for DM management and the exploration order for dynamic multimedia and wireless network applications, proposed in Sects. 6.2 and 6.3, respectively, suitable sets of custom DM manager candidates are selected for each application. This second phase can take up to one week for very complex applications, namely with more than 20 de/allocation phases in the same application.

Next, in the third phase of the DM management methodology we implement the DM manager candidates and exhaustively explore their run-time behavior in the application under study, such as memory footprint, fragmentation, performance and energy consumption. For this purpose, as we have explained in Sect. 6.2.3, we have developed an optimized C++ library that covers all the decisions in our DM design space and enables the construction and profiling of our custom DM managers implementations in a simple way via composition of C++ layers [12]. As a result, the whole implementation space exploration is done automatically using the aforementioned library and additional tools, which execute multiple times the application with the different DM managers to acquire complete run-time profiling information. This third phase takes between 3 and 4 days in case of complex applications.

In the fourth and final phase of the DM management design flow, we use our own tools to automatically evaluate the profiling generated in the previous phase and to determine the final implementation features of the custom DM manager for the application, such as the final number of max block sizes and the number of allowed allocation memory sizes. This fourth phase takes between 2 and 3 days. Overall, the whole flow for custom DM managers requires between 2 and 3 weeks for real-life applications.

6.5 Case Studies and Experimental Results

We have applied the proposed methodology to three realistic case studies that represent different multimedia and network application domains: the first case study is a scheduling algorithm from the network protocol domain, the second one is part of a 3D image reconstruction system and the third one is a 3D rendering system based on scalable meshes.

In the following subsections we briefly describe the behavior of the three case studies and the proposed methodology is applied to design custom managers that minimize their memory footprint. All the results shown are average values after a set of 10 simulations for each application and manager implementation using 10 additional real inputs from the ones used to profile and design the custom DM managers. All the final values in the simulations were very similar (variations of less than 2%).

6.5.1 Method Applied to a Network Scheduling Application

The first case study presented is the Deficit Round Robin (DRR) application taken from the NetBench benchmarking suite [115]. It is a scheduling algorithm implemented in many routers. In fact, variations of the DRR scheduling algorithm were used by Cisco for commercial access point products and by Infineon in its broadband access devices. Using the DRR algorithm the router tries to accomplish a fair scheduling by allowing the same amount of data to be passed and sent from each internal queue. In the DRR algorithm, the scheduler visits each internal non-empty queue, increments the variable "deficit" by the value "quantum" and determines the number of bytes in the packet at the head of the queue. If the variable "deficit" is less than the size of the packet at the head of the queue (i.e., it does not have enough credits at this moment), then the scheduler moves on to service the next queue. If the size of the packet at the head of the queue is less than or equal to the variable "deficit", then that variable is reduced by the number of bytes in the packet and the packet is transmitted on the output port. The scheduler continues this process, starting from the first queue each time a packet is transmitted. If a queue has no more packets, it is destroyed. The arriving packets are queued to the appropriate node; if no such node exists, it is created. Ten real traces of internet network traffic up to 10 MB/sec have been used [92] to run realistic simulations of DRR.

To create our custom DM Manager we have followed our methodology step by step. As a result, in order to define the logical phases of the application and its atomic DM manager, we first profile its DM behavior with our dynamic data structures profiling approach [98]. Then, we apply our DM management design exploration. First, we make the decision in tree A2 (Block sizes) to have many block sizes to prevent internal fragmentation. This is done because the memory blocks requested by the DRR application vary greatly in size (to store packets of different sizes) and if only one block size is used for all the different block sizes requested, the internal fragmentation would be large. Then, in tree A5 (Flexible block size manager) we choose to `split` or `coalesce`, so that every time a memory block with a bigger or smaller size than the current block is requested, the splitting and coalescing mechanisms are invoked. In trees E2 and D2 (When) we choose `always`, thus we try to defragment as soon as fragmentation appears. Then, in trees E1 and D1 (number of max/min block size) we choose `many` (categories) and `not fixed` because we want to get the maximum effect out of coalescing and splitting mechanisms by not limiting the size of these new blocks. After this, in trees B1 (Pool division based on size) and B2 (Pool structure), the simplest pool implementation possible is selected, which is a single pool, because if there are no fixed block sizes, then no real use exists for complex pool structures to achieve a reduced memory footprint. Then, in tree C1 (Fit algorithms), we choose the `exact fit` to avoid as much as possible memory loses in internal fragmentation. Next, in tree A1 (Block structure), we choose the most simple DDT that allows coalescing and splitting, which is a `doubly linked list`. Then, in trees A3 (Block tags) and A4 (Block recorded info), we choose a `header` field to accommodate information about the `size` and `status` of each block to support splitting and coalescing mechanisms. Finally, after taking these decisions following the order described in Sect. 6.3, according to the proposed design flow we can determine those decisions of the final custom DM manager that depend on its particular run-time behavior in the application (e.g., final number of max block sizes) via simulation with our own customizable C++ library and tools [12] (see Sect. 6.4 for more details).

Then, we implement it and compare our custom solution to very well-known state-of-the-art optimized general-purpose managers, namely Lea v2.7.2 [96] and Kingsley [171]. The Lea allocator is one of the best generic managers (in terms of the combination of speed and memory footprint) [27] and several variations of it are integrated in the different distributions of the GNU Linux OS. It is a hybrid DM manager that includes different behaviors for different object sizes. For small objects it uses some kind of quick lists [171], for medium-sized objects it performs approximate best-fit allocation [171] and for large objects it uses dedicated memory (allocated directly with the `mmap()` function). Also, we compare our approach with an optimized version of the Kingsley [171] DM manager that still uses power-of-two segregated-fit lists [171] to achieve fast allocations, but it is optimized for objects larger than 1,024 bytes, which take their memory from a sorted linked list, sacrificing speed for good fit. A similar implementation technique is used in several Windows-based OSes [43, 44].

In addition, we have compared our custom DM solution with a manually designed implementation of region-semantic managers [66] that can be found in some embedded OSes (e.g. [128]).

As Fig. 6.6 shows, our custom DM manager uses less memory than the Lea 2.7.2 (Linux OS), the Kingsley and Region DM managers. This is due to the fact that our custom DM manager does not have fixed sized blocks and tries to coalesce and split as much as possible, which is a better option in dynamic applications with sizes with large variation. Moreover, when large coalesced chunks of memory are not used,

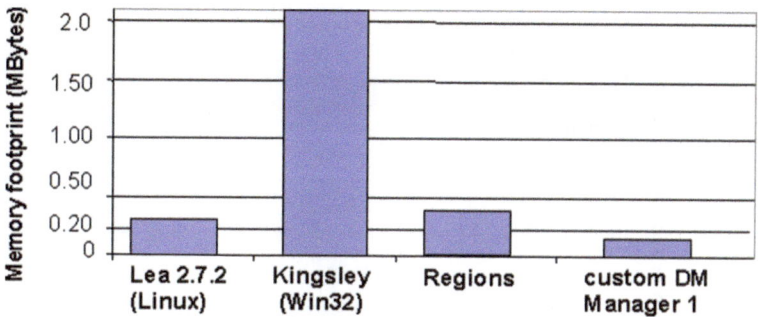

Fig. 6.6 Maximum memory footprint results in the DRR application

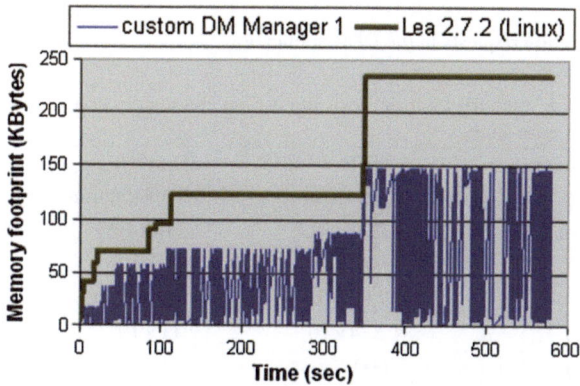

Fig. 6.7 Memory footprint behavior of the Lea DM manager versus the custom DM manager

they are returned back to the system for other applications. On the other hand, Lea and Kingsley create huge freelists of unused blocks (in case they are reused later), they coalesce and split seldom (Lea) or never (Kingsley) and finally, they have fixed sized block sizes. This can be observed in Fig. 6.7, where we show the DM use graphs of the custom and Lea DM managers, and Fig. 6.8, where the DM requests of the application during one run are shown. In the case of the Region DM manager, since no coalescing/splitting mechanisms are applied to reuse memory blocks, more memory footprint than Lea is consumed due to fragmentation.

Concerning the performance of the various DM managers (see Table 6.2 further on for a summary with the results of all the case studies), we can observe that in all cases 600 s is the time used to schedule real traces of 600 s of internet traffic. This means that all the DM managers studied fulfill the real-time requirements of the DRR application. Moreover, our DM manager improves extensively the memory footprint used by the other DM managers, but no extra time overhead is noticed due to its internal maintenance operations.

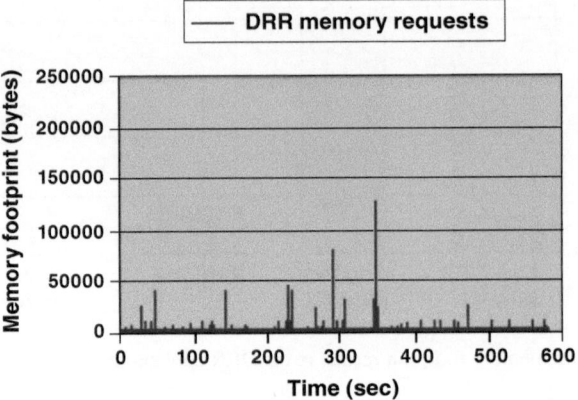

Fig. 6.8 DM allocation requests in the DRR application

Table 6.2 Maximum memory footprint, execution time and energy consumption results of each DM manager in our case studies

Dynamic memorymanagers managers	DRR schedular	3D image reconstruction	3D scalable rendering
Kingsley-Windows (Bytes)	2.09×10^6	2.26×10^6	3.96×10^6
Execution time (s)	600	1.52	6.05
Energy consumption	7.98×10^9	4.04×10^6	11.22×10^6
Lea-Linux (Bytes)	2.34×10^5	2.11×10^6	1.86×10^6
Execution time (s)	600	2.01	8.12
Energy consumption (nJ)	8.18×10^9	4.11×10^6	11.43×10^6
Regions (Bytes)	2.47×10^5	2.08×10^6	2.03×10^6
Execution time (s)	600	1.87	7.03
Energy consumption (nJ)	8.25×10^9	4.01×10^6	11.61×10^6
Obstacks (Bytes)	–	–	1.55×10^6
Execution time (s)	–	–	6.55
Energy consumption (nJ)	–	–	10.93×10^6
Our DM manager (Bytes)	1.48×10^5	1.49×10^6	1.07×10^6
Execution time (s)	600	1.24	6.91
Energy consumption (nJ)	7.11×10^9	3.52×10^6	9.67×10^6

Finally, we have evaluated the total energy consumption of the final embedded system with each of the studied DM managers, using a cycle-accurate ARM-based simulator that includes a complete energy-delay estimation model for 0.13 um [103]. The results indicate that the custom DM manager achieves very good results for energy, when compared to the state-of-the-art DM managers. This is due to the fact that most of the memory accesses performed internally by the managers to their complex management structures are not required in the custom manager, which uses

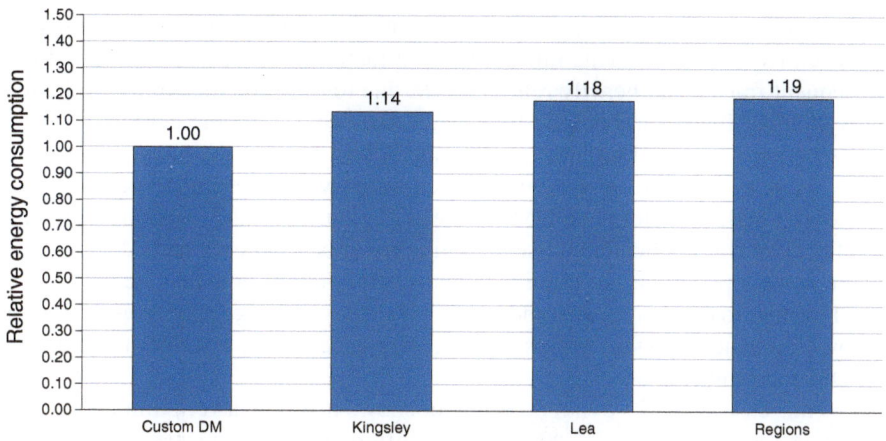

Fig. 6.9 Relative energy improvement of the custom DM manager in the DRR application

simpler internal data structures optimized for the target application. Thus, the custom DM manager reduces by 12, 15 and 16 % the energy consumption values of Kingsley, Lea and Regions, respectively (Fig. 6.9). It is important to mention that even though Kingsley has a smaller amount of DM management accesses, since it does not perform splitting or coalescing operations, it suffers from a large memory footprint penalty. This translates into very expensive memory accesses because bigger memories need to be used.

Consequently, for dynamic network applications like DRR, our methodology allows to design very customized DM managers that exhibit less fragmentation than Lea, Regions or Kingsley and thus require less memory. Moreover, since this decrease in memory footprint is combined with a simpler internal management of DM, the final system consumes less energy as well.

6.5.2 Methodology Applied to a 3D Image Reconstruction System

The second case study forms one of the sub-algorithms of a 3D reconstruction application [138] that works like 3D perception in living beings, where the relative displacement between several 2D projections is used to reconstruct the 3rd dimension. The software module used as our case study application is one of the basic building blocks in many current 3D vision algorithms: *feature selection and matching*. It has been extracted from the original code of the 3D image reconstruction system (see [160] for the full code of the algorithm with 1.75 million lines of high level C++), and creates the mathematical abstraction from the related frames that is used in the global algorithm. It still involves 600,000 lines of C++ code, which demonstrates the complexity of the applications that we can deal with our approach and the necessity of tool support for the analysis and code generation/exploration phases in our overall approach (see Sect. 6.4). This implementation matches corners [138] detected in 2 subsequent frames and heavily relies on DM due to the unpredictability of the features of input images at compile-time (e.g., number of possible corners to match varies on each image). Furthermore, the operations done on the images are particularly memory intensive. For example, each image with a resolution of 640 × 480 uses over 1 MB. Therefore, the DM overhead (e.g., internal and external fragmentation [171]) of this application must be minimized to be usable for embedded devices where more applications are running concurrently. Finally, note that the accesses of the algorithm to the images are randomized, thus classic image access optimizations as row-dominated accesses versus column-wise accesses are not relevant to reduce the DM footprint further.

For this case study, its dynamic behavior shows that only a very limited range of data type sizes are used in it [98], namely 8 different allocation sizes are requested. In addition, most of these allocated sizes are relatively small (i.e., between 32 or 16,384 Bytes) and only very few blocks are much bigger (e.g., 163 KB). Furthermore, we see that most of the data types interact with each other and are alive almost all the execution time of the application. Within this context, we apply our methodology and using the order provided in Sect. 6.3 we try to minimize the DM footprint wastage (e.g., fragmentation, overhead in the headers, etc.) of this application. As a result, we obtain a final solution that consists of a custom DM manager with 4 separated pools or regions for the relevant sizes in the application. The first pool is used for the smallest allocation size requested in the application, i.e., 32 bytes. The second pool allows allocations of sizes between 756 bytes and 1,024 bytes. Then, the third pool is used for allocation requests of 16,384 bytes. Finally, the fourth pool is used for big allocation requests blocks (e.g., 163 or 265 KB). The pool for the smallest size has its blocks in a single linked list because it does not need to coalesce or split since only one block size can be requested in it. The rest of the pools include doubly linked lists of free blocks with headers that contain the size of each respective block and information about their current state (i.e., in use or free). These mechanisms efficiently support immediate coalescing and splitting inside these pools, which minimizes both internal and external fragmentation in the custom DM manager designed with our methodology.

In this case study we have compared our solution with region-semantic managers [66, 171]. Also, we have tested our manager with the same optimized versions of Kingsley and Lea used in the previous example (i.e., DRR) since these are the types of DM managers found in embedded systems. The memory footprint results obtained are depicted in Fig. 6.10.

These results show that the values obtained with the DM manager designed using the proposed methodology obtains significant improvements in memory footprint compared to the manually designed implementation of a Region manager (28.47 %), Lea (29.46 %) and the optimized version of Kingsley (33.01 %). These results are because our custom DM manager is able to minimize the fragmentation of the system in two ways. First, because its design and behavior varies according to the different block sizes requested. Second, in pools where a range of block sizes requests are allowed, it uses immediate coalescing and splitting services to reduce both internal and external fragmentation. In the region managers, the blocks sizes of each different region are fixed to one block size and when blocks of several sizes are used, this creates internal fragmentation. In Lea, the range of block sizes allowed does not fit exactly the ones used in the applications and the mixed de/allocation sizes of few block sizes produce a waste of the lists of sizes not used in the system. Furthermore, the frequent use of splitting/coalescing mechanisms in Lea creates an additional overhead in execution time compared to the Region manager, which has better adjusted allocation sizes to those used in the application. Finally, in Kingsley, the coalescing/splitting mechanisms are applied, but an initial boundary memory is reserved and distributed among the different lists for sizes. In this case, since only a limited amount of sizes is used, some of the "bins" (or pools of DM blocks in Kingsley) [171] are underused.

In addition, the final embedded system implementation using our custom DM manager achieves better energy results than the implementations using general-purpose DM managers. In this case study, our DM manager employs less management accesses to DM blocks and memory footprint than any other manager. Thus, our DM manager enables overall energy consumption savings of 10 % with respect to Regions, 11 % over Kingsley and 14 % over Lea (Fig. 6.11).

6.5.3 Methodology Applied to a 3D Video Rendering System

The third case study is the 3D rendering module [172] of a whole 3D video system application. This module belongs to the category of 3D algorithms with scalable meshes [104] that adapt the quality of each object displayed on the screen according to the position of the user watching at them at each moment in time (e.g., Quality of Service systems [136]). Therefore, the objects are represented by vertices and faces (or triangles) that need to be dynamically stored due to the uncertainty at compile time of the features of the objects to render (i.e., number and resolution). First, those vertices are traversed in the first three phases of the whole visualization process,

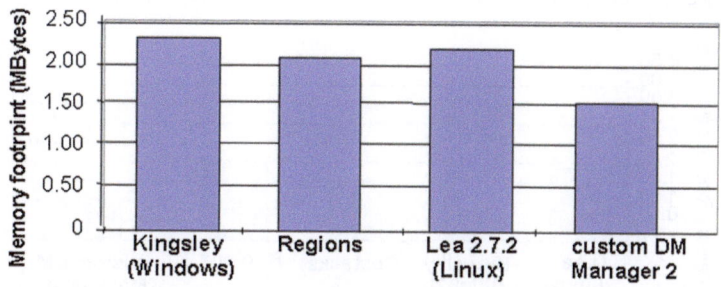

Fig. 6.10 Maximum memory footprint results in the 3D reconstruction application

i.e., model-view transformation, lighting process and canonical view transformation [172]. Finally, the system processes the faces of the objects in the next three phases (i.e., clipping, view-port mapping and rasterization [172]) of the visualization process to show the final object with the appropriate resolution on the screen. According to our experiments, this application closely resembles the DM behavior of the MPEG4 Visual Texture deCoder (VTC) implementation in the standard [124].

In this case, we have compared our custom manager with Lea v2.7.2, the optimized version of Kingsley, the Region manager and due to its particular behavior with phases where intermediate values are built in a sequential way and are finally destroyed at the end of each phase (for the phases that handle vertices), we have also used Obstacks [171]. Obstacks is a well-known custom DM manager optimized for applications with such stack-like behavior. E.g., Obstacks is used internally by *gcc*, the GNU Compiler.

As Fig. 6.12 shows, Lea and the Region manager obtain better results in memory footprint than the Kingsley DM manager. Also, due to the stack-like behavior of the application in the phases that handle triangles, Obstacks achieves even better results than Lea and region managers in memory footprint. However, the custom

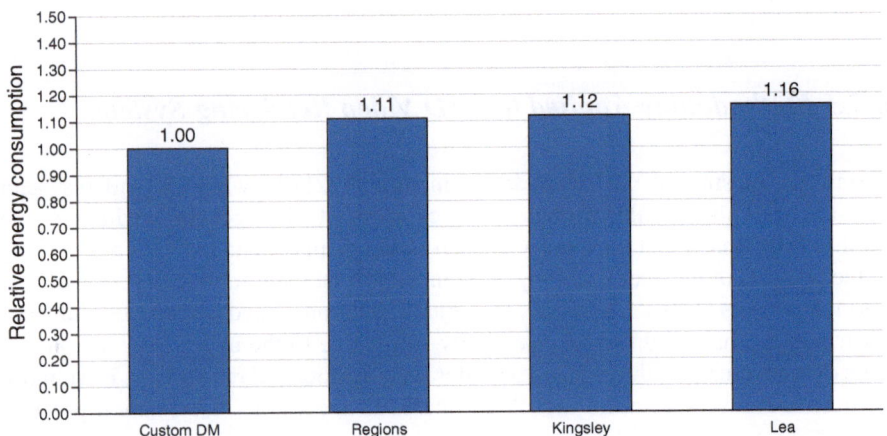

Fig. 6.11 Relative energy improvement of the custom DM manager in the 3D reconstruction application

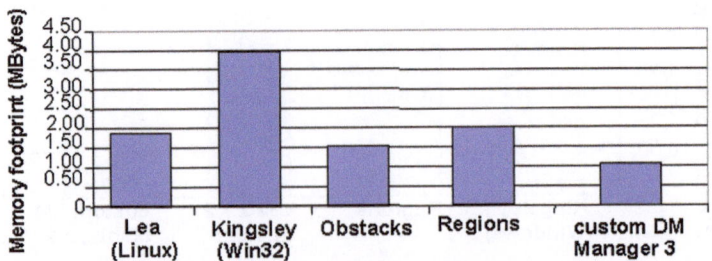

Fig. 6.12 Maximum memory footprint results in the 3D rendering application

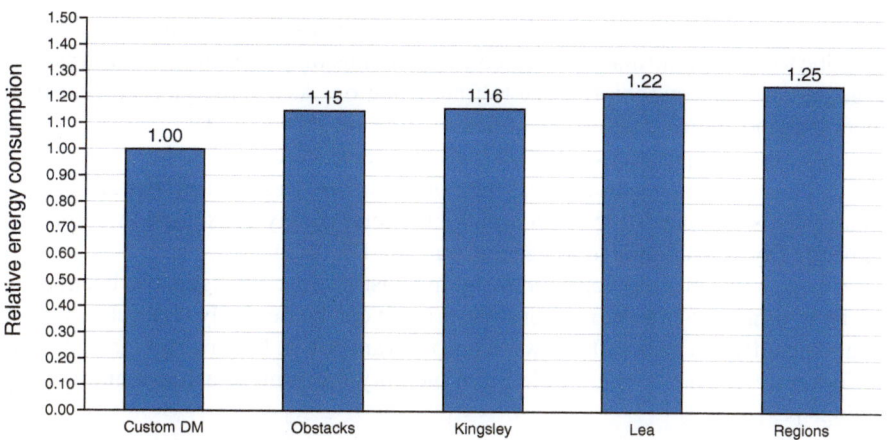

Fig. 6.13 Relative energy improvement of the custom DM manager in the 3D rendering application

manager designed with our methodology improves further the memory footprint values obtained by Obstacks. The fact is that the optimized behavior of Obstacks cannot be exploited in the final phases of the rendering process because the faces are used all independently in a disordered pattern and they are freed separately. Thus, Obstacks suffers from a high penalty in memory footprint due to fragmentation because it has not got a suitable maintenance structure of the DM blocks for such DM de-allocation behavior.

From an energy point of view, our custom DM manager also improves the results obtained with the studied general-purpose managers in the ARM-based simulator. In a similar way as in the previous case studies, our DM manager requires less memory management accesses than Lea and Regions, thus improving the global energy figures of the final embedded system by 18, 20 %, respectively. In the case of Obstacks and Kingsley, they produce less memory accesses than our custom DM manager due their optimized management of DM blocks for performance, but their larger consumption of memory footprint finally enables 13 and 14 % overall savings in energy consumption for our custom DM manager, respectively (Fig. 6.13).

Finally, to evaluate the design process with our methodology, our proposed design and implementation flow of the final custom DM managers for each case study took us two weeks. Also, as Table 6.2 shows, these DM managers achieve the least memory footprint values with only a 10 % overhead (on average) over the execution time of the fastest general-purpose DM manager observed in these case studies, i.e., Kingsley. Moreover, the decrease in performance is not relevant since the custom DM managers preserve the real-time behavior required by the applications. Thus, the user will not notice any difference. In addition, the proposed custom DM managers include optimized internal DM management organizations of memory blocks for each studied application, which usually produce a decrease in memory accesses compared to state-of-the-art general-purpose managers, such as Lea or Regions, that are designed for a wide range of memory requests and DM behavior patterns. Only

Kingsley and Obstacks produce less memory accesses than the custom DM manager due to their performance-oriented designs, thus outperforming it, but wasting a large part of memory footprint due to fragmentation. As a consequence, these managers demand larger on-chip memories to store the dynamic data and much more energy is required per access [103], which counteracts the possible improvements in energy consumption because of less memory accesses. In summary, our custom DM managers also reduce by 15 % on average the total energy consumption of the final embedded system compared to general-purpose DM managers.

Although our methodology for designing custom DM managers has been driven by the minimization of the memory footprint, it can be perfectly retargeted towards achieving different trade-offs between any relevant design factors, such as improving performance or consuming a little more memory footprint to achieve more energy savings [13].

6.6 Conclusions

Nomadic embedded devices have improved their capabilities in the last years making feasible to manage very complex and dynamic multimedia applications. Such applications have grown lately in complexity and to port them to the final embedded systems, new design methodologies must be available to efficiently use the memory present in these very limited embedded systems. In this chapter we have presented a systematic methodology that defines and explores the dynamic memory management design space of relevant decisions, in order to design custom dynamic memory managers with a reduced memory footprint for dynamic multimedia and wireless network applications. Our results in real applications show significant improvements in memory footprint over state-of-the-art general-purpose and manually optimized custom DM managers, incurring only a small overhead in execution time over the fastest of them.

Chapter 7
Systematic Placement of Dynamic Objects Across Heterogeneous Memory Hierarchies

In the previous chapters we have explained how to improve different aspects of the memory subsystem when dynamic memory is used in an embedded system. In particular, we explained how to design efficient custom dynamic memory managers to serve the dynamic memory requests of the applications. Thereby, Chap. 6 shows how to deal with the allocation problem, that is, finding memory blocks to satisfy the application requests with the least possible additional computation and minimizing the total memory footprint. The first part is achieved via efficient algorithms that require a limited amount of operations to locate the appropriate blocks. The second part is achieved both reducing the additional footprint of the data structures used internally by the DMM itself and minimizing the amount of fragmentation.

However, embedded systems frequently have complex memory organizations that include several memories of different characteristics that are directly addressable, that is, they form part of the architecture and are explicitly present in the memory map that system programmers can use. The reason is that, in those systems, cache memories are not always the most efficient option [19]: Basically, cache memories need additional area to storage the line tags and accommodate the controller logic; they also consume additional energy in the control logic and tag comparators. This situation is in stark contrast to the flat view offered to programmers in desktop systems. As a consequence, along the years different groups have proposed various techniques to exploit the presence of SRAM memories (commonly known as "scratchpad" memories) for the processing of static data elements [2, 4, 37, 50, 67, 86, 87, 102, 156, 167, 168, 176]. One of their main goals was to improve the processing of big blocks of static data, such as frame buffers in video processing applications.

The use of dynamic memory introduces several new challenges. First, software techniques for the placement of static data over scratchpad memories work on concrete data instances of known size, but the use of dynamic memory precludes knowing the number of instances and their size at design-time. In fact, a complexity jump is present from working on static data instances declared at specific locations in the source code versus dealing with multiple instances created at the same point (the point where the call to the memory allocator is made). Consider, for instance, the case of an application that allocates a variable number of arrays of different size to

© Springer International Publishing Switzerland 2015
D. Atienza Alonso et al., *Dynamic Memory Management for Embedded Systems*,
DOI 10.1007/978-3-319-10572-7_7

accommodate a group of specific inputs. Moreover, these techniques are frequently limited to the use of a single specific scratchpad memory. Second, cache memories are based on the principle of locality [134] to amortize the cost of transferring data from and to the main memory over several consecutive hits. However, logically-adjacent linked nodes are not necessarily stored in consecutive memory addresses (nodes may be added and deleted at any position in a dynamically linked structure), breaking the spatial locality assumption, and, in some structures such as trees, the path taken can be very different from one traversal to the next one, thus destroying temporal locality as well.

Therefore, the use of dynamic memory in embedded systems with complex memory organizations brings a new flavor to an old challenge: Deciding the most appropriate physical location for each new memory allocation in such a way that the most accessed data reside in the most efficient memories. However, classical DMM solutions, originally devised to deal with all sorts of applications on desktop and server computers, take good care of finding a free memory block, but typically exploit *any* free memory block, for the requests as they arrive; as long as they are adequate for the requested allocation size, all blocks are considered the same regardless of their physical position in the system memory. We refer to this as "the placement problem" (Fig 7.1).

In this chapter, we present a proposal to tackle the placement problem in the context of embedded systems with complex memory organizations. This proposal uses an exclusive memory model, where the memories closer to the processor do not necessarily duplicate data from the bigger and slower memories. In comparison, cache hierarchies tend to favor an inclusive memory model wherein the smaller levels keep a subset of the data contained in the bigger ones. As a side effect, this also enables its utilization with lightweight platforms that only have small on-chip software-controlled memories, where data movements are not applicable. Additionally, it is also compatible with the techniques presented in Chap. 4 for the reorganization of

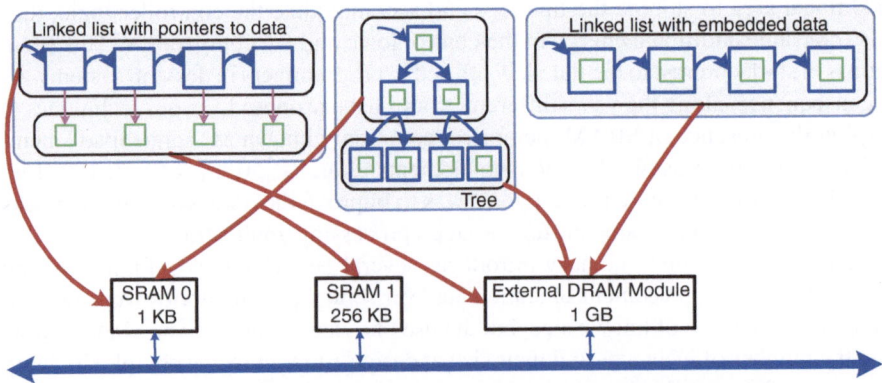

Fig. 7.1 In this chapter we propose a methodology to map the DDTs of an application into the elements of the memory subsystem so that the most accessed DDTs are placed on the most efficient resources

DDTs according to their access patterns. For example, after splitting of inter-node pointers and data, this proposal can distinguish between both and tightly pack the pointers into an efficient memory to improve traversals, while the data elements, which may be much bigger, are stored in a different memory and accessed only when really needed.

The overall methodology is proposed in Sect. 7.1. The detailed approach is discussed in Sect. 7.2 and the experimental results are shown in Sect. 7.3. Finally, Sect. 7.4 contains the comparisons to the related work.

Through the rest of this chapter, we shall use the term *heap* to denote the address range, *dynamic memory manager* (DMM) for the algorithms and *pool* for the algorithms and the address range as a whole.

7.1 Methodology for Placement of Dynamic Data

The methodology proposed in this chapter is based on the principles presented along this book. In essence, it can be seen as an optimization step that is applied after the application has been characterized and the optimizations at the DDT level from Chap. 4 applied. Then, the final DDTs of the application are analyzed and grouped according to their characteristics. Each group is treated with the techniques for DMM presented in Chap. 6, as if the DDTs in that group where the only ones present in the application. Finally, each pool produced by DMM is placed into the modules of the memory subsystem according to the type of accesses and the memory properties and taking advantage of the fact that modern DMMs can split a pool over several memory modules, breaking the constraint of mapping each pool to a single one.

The inputs to the placement methodology are the characterization of the application behavior and its most relevant dynamic data types, and the template for the targeted memory organization (Fig. 7.2). As output, it generates a description of the memory pools (which includes the algorithms required to manage their heaps), their association (address assignment) with physical memory resources and a list of the DDTs they will lodge (Fig. 7.3).

The methodology is divided in the following steps:

1. **Instrumentation** The source code of the application is instrumented as presented in Chap. 3. In this particular case, the same instrumentation will be reused in the final deployment step to implement the different dynamic memory managers.
2. **Profiling and analysis** The application is profiled with typical input scenarios and a log file is generated. We perform then an analysis of the log file and generate a metadata representation of the application DDTs as explained in Chap. 3.
3. **Group creation** The application DDTs are classified according to their characteristics (memory footprint evolution, frequency of accesses per byte, etc.) to adjust the trade-off between creating a pool for every DDT, and putting all of them together in a single pool. Conversely, DDTs that could benefit from being placed

in different memory resources are separated into different groups. We introduce the concepts of liveness and exploitation ratio to guide this step.

4. **Definition of pool algorithms** With the list of the DDTs that will be allocated in each pool, we use the methods presented in Chap. 6 to determine the algorithms and internal data structures of each of them.

5. **Mapping into memory resources** We use the description of the memory subsystem to map all the pools into the memory modules of the platform, assigning physical addresses to each pool. This phase is quite straightforward (under our assumptions): The algorithm employed is an instance of the classic fractional knapsack problem. The output of this step, which is the output of the whole methodology, is a list of memory address ranges for each of the pools.

6. **Simulation and evaluation** We have created a complete simulation environment to evaluate the mapping solutions before deployment into the final platform. Alternatively, if the exploration is performed before the platform is finalized, the results obtained with it can be used to steer the design or selection of the platform.

7. **Deployment** Finally, the description of the pools needed to perform the allocation and placement of the application DDTs is generated as metadata that will be distributed with the application. The size of the metadata should not constitute a considerable overhead on the size of the deployed application. In order to attain maximum flexibility, we propose to employ at run-time a factory of DM managers in combination with the strategy design pattern [64] to construct the required memory managers according to their template description.

The most novel characteristic of this methodology is the grouping step. Solving the placement problem means placing all the instances of all the DDTs of the application into the memory elements, but this can be done at several abstraction levels. For instance, it would be very difficult to solve the problem at the level of each different instance because their numbers are unknown. The opposite direction would be to group all the DDTs into a single pool and allocate memory as needed during runtime; however, this would be quite close to the standard behavior of DMM without placement constraints. As an intermediate solution, we propose to group DDTs with similar characteristics into one pool, and assign memory resources to the whole pool

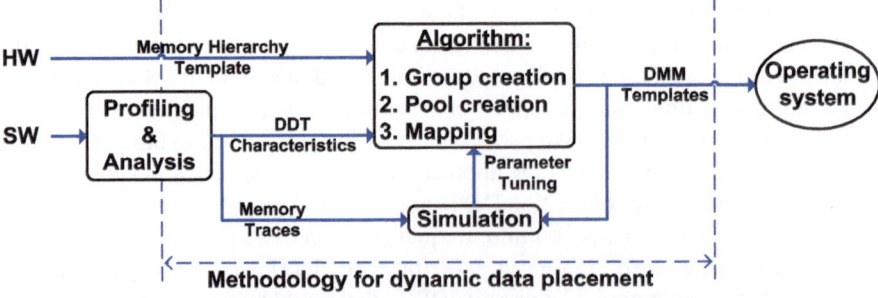

Fig. 7.2 Overview of the methodology for dynamic data placement

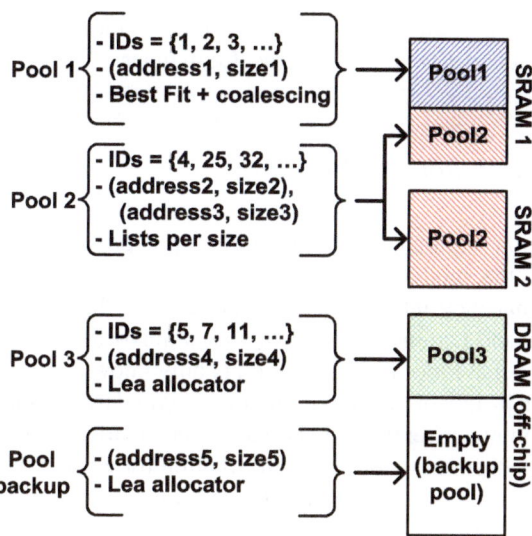

Fig. 7.3 Outcome of the methodology for dynamic data placement: pool description and placement on memory resources

so that all the instances of these DDTs are placed in the resources of the pool indistinctly. The difficulty of grouping lies then on identifying DDTs whose instances have similar access frequencies and patterns, so that any of these instances would achieve a similar exploitation of the assigned resources.[1] However, as the methodology presented here does not execute data movements across elements in the memory hierarchy, there is a second trade-off between leaving resources underused when there are not enough alive instances of the DDTs assigned to the corresponding group, and allowing instances from less accessed DDTs to use better memory resources when there is some space left in them. If all the DDTs are grouped together in a single pool, we risk that newer instances of the demanding DDTs will have to use less efficient resources because the better ones are already occupied—perhaps, by much less accessed instances. The grouping mechanism tries to compensate for it by considering together DDTs that have complementary footprint demands along time, so that "valleys" from ones compensate for "peaks" of the others and the overall exploitation ratio of the memory resources is kept as high as possible during the whole execution time.

Finally, the methodology is completely modular as each step interacts with the rest through DDT, group and pool abstractions; hence, any step can be improved independently. As an example, the mapping step could be integrated with the run-time loader of the operating system to take into account resource availability when the

---

[1] It is possible that each of the instances of one DDT has very different access characteristics. A hypothetical mechanism to differentiate between them would be an interesting basis for future work.

application is started, optimize for different implementations of the same platform or tackle with system degradation in a graceful manner.

7.1.1 Instrumentation and Profiling

The instrumentation required for the profiling step is presented in Chap. 3. In summary, it requires the modification of those classes whose instances are created dynamically so that they inherit from a generic class that overloads the `new` and `delete` operators and parameterizes them with a univocal identifier for each DDT. Even if the exact behavior of the application is only known at run-time, it is usually possible to gain an insight of the most common cases and trigger the application using representative inputs to cover them. This enables the identification of different system scenarios and the construction of a specific solution for each of them. The operating system of the embedded device can choose at run-time the adequate solution according to the actual scenario detected at each moment making the methodology adaptive to run-time events and variations in behavior, as explained in [68].

The modifications needed for deployment into the target platform are exactly the same. Therefore, if application development is planned with that in mind, the overhead imposed by our methodology on the designers is minimal given the significant improvements that can be attained. Moreover, similarly to how the profiling methods explained in Chap. 3 are valid for applications written in languages other than C++, deployment is also possible as long as the execution environment offers mechanisms to implement and access the generated dynamic memory managers.

7.1.2 Analysis

The analysis step in this methodology is based on the one presented in Chap. 3. In addition to the information already mentioned there (maximum number of instances that are currently alive, maximum footprint, number of accesses per DDT instance, allocation and de-allocation sequence, etc.), the analyses step gives the designer the following information for each DDT:

- **Frequency of accesses per byte (FPB)** Number of accesses to a data object divided by its peak footprint, calculated for all the instances of the DDT together.
- **"Liveness,"** that is, the evolution of the DDT footprint along execution time as a list of footprint variations, which we designate as the DDT behavior. The liveness of a DDT is calculated by tracking all the calls to the allocation functions recorded in the profiling log. The behavior of each DDT has one entry for each instant when the footprint of that DDT changes. Each entry includes information on the number of accesses to the instances of the DDT since the previous one. This organization avoids handling a single list with all the allocation events in the application that

would have empty entries for the DDTs that do not have footprint variations on a given instant. This helps to reduce the memory footprint of the analysis tool itself.

The analysis tool can distinguish between instances of the same DDT created with different sizes (e.g., multiple vector instances of varying sizes dynamically allocated at a given source code location). We use this as a specialization of the DDT concept through the rest of this work, that is, we consider these different-size instances as distinct DDTs.

7.1.3 Group Creation

Grouping is the main step of the proposed methodology. It evaluates the trade-off between assigning each DDT to a separate group and assigning all them to a single group. As each group is transformed into a pool during the next step, two DDTs assigned to different groups are kept in separate pools and, if possible, they will be placed in different memory resources. Similarly, two DDTs with complementary characteristics that are included in the same group will be managed in the same pool; thus, their instances will be placed on the resources assigned to the pool indistinctly. The difficulty of grouping lies then on identifying DDTs whose instances have similar access frequencies and patterns, so that any of these instances would achieve a similar exploitation of the assigned resources. However, there is a second aspect in this trade-off that involves balancing between leaving resources underused when there are not enough alive instances of the DDTs assigned to the corresponding group, and allowing instances from less accessed DDTs to use better memory resources when there is some space left in them. If all the DDTs are grouped together in a single pool, it is possible that newer instances of the demanding DDTs will have to use less efficient resources because the better ones are already full. The grouping algorithm tries to compensate for these effects by considering together DDTs that have complementary footprint demands along time, so that "valleys" in the footprint of some DDTs compensate for "peaks" of the others and the overall exploitation ratio of the memory resources is kept as high as possible during the whole execution time. These factors make the grouping step the most important one in the methodology.

The grouping of DDTs is a hard problem whose optimal solution cannot be found with a greedy algorithm. However, the computational complexity of an optimal solver may be unaffordable for real-life cases. In this chapter we propose the use of a greedy algorithm that offers several *"knobs"* to steer the grouping process and adapt it to the specific features of the application under development. Even if this option is not optimal, the performance of the solutions generated for the case studies of this chapter improve significantly on the ones that can be achieved with caches.

7.1.4 Definition of Pool Algorithms

We use the concept of pool to represent one or several address ranges (*heaps*) reserved for the allocation of DDTs and the data structures and algorithms needed to manage them. For every group from the previous step, a pool that inherits its list of DDTs is generated. Considerations like the degree of block coalescing and splitting, choice of fit algorithms, internal and external fragmentation, and number of accesses to internal data structures and their memory overhead can be evaluated as explained in Chap. 6.

The result of this step is a list of pools ordered by their FPBs, and the description of the chosen algorithms in a form of metadata that can be used to build the memory managers at run-time.

7.1.5 Mapping into Memory Resources

The input to the mapping step is the ordered list of pools with their internal structure and a description of the memory subsystem of the platform. The algorithm produces a placement of the pools into the available memory resources. The result of this step is a list of pools, where each pool is annotated with a list of address ranges that represent its placement into memory resources. The computational complexity of this algorithm is in the order of $O(n)$, being n the number of pools.

In our approach to the problem we take three important decisions. First, we allow to split pools over several memory resources even if they are not mapped on consecutive addresses. This is technically feasible with modern DMMs at the expense of perhaps a slightly larger fragmentation (however, if the pools are mapped into memory blocks with consecutive memory addresses, this overhead disappears because the blocks become a single entity at a logical level). This reduces the mapping problem to an instance of the fractional knapsack problem [31], enabling the utilization of a greedy algorithm to solve it optimally. Therefore, the ability to split pools over several memory modules allows us to obtain better solutions while using an efficient algorithm that could be moved in the future to a run-time phase.

The second decision is that we focus on memory organizations where only one memory module can be accessed at the same clock cycle. It is possible to consider situations where the processor has simultaneous access to several independent buses; however, a greedy algorithm would not be enough for that new problem and a more complex approach such as the one presented by [151] would be needed. On the contrary, our algorithm can consider the effects of mapping DDTs with interleaved accesses into multi-bank DRAMs to minimize the number of row changes in the DRAM banks.

Finally, the third decision is the assumption that all the instances created in the pool have the same FPB. It could be interesting to study more complex access patterns in the future.

7.1.5.1 The Algorithm

Algorithm 7.1 presents the pseudo-code for the mapping algorithm. First, the memory modules are ordered by the targeted cost function, either energy or latency. Then, as many pools as possible are mapped into the current module until it is full. If a pool is bigger than the space left in the module, it is split and the remaining size is assigned to the next memory module. Splitting a pool can introduce some fragmentation; thus, the designer can specify a small size increase in those cases (lines 10–12). When a pool is split, the remaining size is rounded up to the size of an instance if the it contains only one DDT; otherwise, the parameter PoolSizeIncreaseFactorForSplitPools is used.[2]

Algorithm 7.1 Mapping

```
 1: function Mapping(List of pools, List of memory blocks, Cost function)
 2:    Order memory blocks using the cost function (energy / access time)
 3:    For each pool in the list of pools do
 4:        Multiply pool size by PoolSizeIncreaseFactor
 5:        While the pool is not fully mapped do
 6:            Select the first block in the list of available blocks
 7:            If the remaining size of the pool > available space in the block
           then
 8:                Assign the available space in the block to the pool
 9:                Remove the block from the list of memory blocks
10:                If the pool has only one DDT then
11:                    Round up pool's remaining size to a multiple of the DDT
           instance size
12:                Else Increase the pool's remaining size by
           PoolSizeIncreaseFactorForSplitPools
13:            Else Assign the rest of the pool to the block
14:                If block is DRAM and SpreadPoolsInDRAMBanks then
15:                    Assign the whole DRAM page
16:                Else Reduce available space in the block
17:                    If the available space in the block < MinMemoryLeftOver
           then
18:                        Assign everything to the pool and remove block from
           the list of memory blocks
19:    return (blocks, pools)
20: end function
```

---

[2] In extreme cases of platforms with many small memory blocks, this adjustment can lead to an unbounded increase of the pool size. To avoid this effect, the designer can use a smaller value or the algorithm could be changed to increase the size only the first time that the pool is split.

7.1.5.2 Algorithm Parameters

The mapping algorithm offers several parameters that the designer can use to tune the solution to the application:

1. **MinMemoryLeftOver** Minimum acceptable size for any pool fragment (the default is 8 bytes). Some DMM algorithms may impose a bigger minimum block size.
2. **Backup pool** Several factors, such as differences in the current inputs with respect to the inputs used during profiling, may increase the actual memory requirements during execution beyond those obtained during the profiling phase. A backup pool can be created to host all the DDT instances that cannot be allocated in their corresponding pools. Accesses to these instances will typically carry a penalty, as the backup pool is usually located in the main DRAM. Hence, although the backup pool adds robustness to the system, it is not an excuse to avoid a detailed profiling.
3. **PoolSizeIncreaseFactor** The designer can adjust the amount of memory assigned to each pool. This allows balancing between covering the worst-case and reducing the amount of space wasted during the most common cases. Our experiments show that the liveness analysis packs enough DDTs in each group so that the higher exploitation ratio of the pool avoids wasting too much memory and a value of 1.0 gives usually the best results.
4. **PoolSizeIncreaseFactorForSplitPools** Splitting a pool over several memory modules (with non-contiguous address ranges) may increase its fragmentation and, hence, its footprint. The default value is 1.0, but a value as high as 1.3 was needed in some of our internal experiments.
5. **MappingGoal** Memory modules can be ordered according either to energy or access delay cost. Although in the following case studies we consider memory technologies that improve both simultaneously (increasing the area cost per bit), either goal may be specified.
6. **SpreadPoolsInDRAMBanks** This parameter produces a simple spreading of pools over DRAM banks according to FPB ordering. The idea is that accesses to the instances of the DDTs have a higher probability of hitting different DRAM banks, thus reducing the number of costly page misses. If there are more pools than DRAM banks, the remaining ones are all placed in the last bank (better options are conceivable taking into account the interactions between accesses to the pools).

7.1.6 Simulation and Evaluation

The mapping step produces the information needed to execute the application on the final system. However, if the actual platform is not yet available, maybe because the design is still at an early stage, a memory subsystem simulator can be used to get an estimation of the placement performance. The designer can use this memory subsystem simulator to judge the generated placement, check if the design constraints are met and evaluate different memory organizations before selecting the final hardware. To this end, we implemented a simulator that, using a complete simulation model of the memory subsystem, reproduces the behavior of the profiled application (through the memory traces) to compute the energy and number of cycles required for each access.

The inputs to the simulator are the pool mappings, the log obtained during profiling and the template of the memory organization with annotated costs that was used during the mapping step. The memory subsystem template can include any combination of the following elements:

1. **Multi-level bus structures** The memory subsystems can be connected to the processor through a common bus, or through a multi-level hierarchy of buses with bridges between them.
2. **Static RAMs** The simulator supports SRAMs of any size, parameterized with their energy and latency cost per access. As an example, scratchpads are usually implemented as SRAMs.
3. **Dynamic RAMs** Main memory is usually implemented as DRAM due to its higher density and lower cost per bit. A fundamental characteristic of DRAMs is that they are organized in banks and pages (rows). Each bank can have only one page active at a time; switching to a different page involves saving the currently open page to the main DRAM matrix and reading the new page. Therefore, the cost of an access depends on the currently active page for the corresponding bank. Additionally, DRAMs are based on capacitors that need to be "refreshed" in order to keep their charges. As a result, the DRAM becomes inaccessible periodically (every few microseconds) as the controller refreshes one row each time. The simulator supports two types of DRAMs: SDRAM and LPDDR-X-SDRAM. For both cases, it uses the appropriate physical parameters and the state of all the banks (which page is open on each of them) to calculate the cost of every access, the energy cost of driving the memory pins and the background energy consumed. These calculations are performed according to the rules for state transitions defined by the JEDEC association [81] and the manufacturer's datasheets [119, 120].

   (a) **Mobile SDRAM** Single data rate, low power (mobile) version. Multiple consecutive read or write (not interleaved) accesses can be issued to random addresses in different banks and the memory outputs one word of data per cycle as long as all the involved banks are active and no page misses happen. This technology allows burst sizes of 1, 2, 4, 8 or full-page.
   (b) **LPDDR-X-SDRAM** This low-power version of the double data rate (DDR) standard is modeled after the JEDEC's specifications [81]. This technology transfers two data words per cycle, using both the rising and falling edges of the clock. Consecutive bursts may be executed to any number of active banks without extra latency as long as the appropriate pages are selected.

4. **Cache memories** Every DRAM module can have an associated cache hierarchy. Cache memories may be modeled using several parameters: size, associativity (from direct-mapped up to 16-ways), line length (4, 8 or 16 words), replacement policy (random or Least-Recently-Used - LRU), cached address range and energy and latency costs per access.

Due to the fact that the original application is commonly profiled using a different memory organization than the one used for mapping and simulation, especially in the case of platform exploration, the simulator has to convert the addresses from the original profiling log into the corresponding addresses in the new memory organization, tracking the allocation and placement of DDT instances. In the simulator version used in our experiments, we assumed an ideal DMM manager that has no overhead and no fragmentation. However, it can be easily extended to use a library of DMMs as explained in Sect. 7.1.7, taking into account the differences between executing the code in a real platform and simulating its behavior through memory traces.

7.1.7 Final Deployment

We propose two different alternatives to prepare the application for execution on the final system. The first one is to use the library for the construction of dynamic memory managers presented in Chap. 6. However, this method is based on the use of *mixins* and hence, the resulting DMMs are fixed at design time. The second alternative is to use a combination of the factory and strategy design patterns [64] to construct the dynamic memory managers. In this way, a generic factory class and library components (strategy implementations) can be used to build the DMMs at run-time according to their template description. This would also allow for the implementation of several solutions each suited for a different system scenario.

The final deployment of the application comprises the application's compiled code, including the mixins-based library of DMMs, or, alternatively, the application's compiled code, metadata with the description of the pools, the library of DMM components and the code for the factory object.

7.1.8 Suitability and Current Limitations

The technique that we have presented is best suited in its current form for applications that use DDTs in phases or traverse data structures accessing only a small amount of data at each node, and whose DDTs have very different FPBs. These cases may hinder the performance of hardware caches as they have to move more data around the memory hierarchy than is really needed, possibly evicting very accessed objects with seldom accessed ones, but they offer a chance for memory resources reutilization that can be exploited by an exclusive memory assignment like ours. As an example, the traversal of an structure could force a cache to move complete nodes back and forth, with an increasing waste of energy as the size of the cache lines increases. Splitting of data structures may allow to pack the pointers of many nodes tightly and access only the data of the nodes that are actually needed during the traversal. This effect is particularly beneficial with our methodology because the pool containing the pointers is guaranteed to be always in the correct memory module, independently of which other data accesses are performed by the applications.

On the contrary, for DDTs with many instances where some of them get many more accesses than others, this proposal may suffer from the inability to distinguish between instances with very different FPBs.

Finally, our methodology leaves the least accessed DDTs in the most distant memory modules; this is appropriate for data streams that are processed sequentially and without a high data reuse, or for small data elements that are seldom accessed. However, for other access patterns to big structures, it may be convenient to supplement our approach with a small HW cache or with SW techniques for the movement of array sections such as array tiling or blocking.

7.2 Detailed Approach for Group Creation

We will now discuss more in-depth the different aspects of the group creation task.

7.2.1 Goals

The grouping process has three main goals:

1. Decreasing the complexity of individually placing every instance of each DDT into the memory subsystem, working instead with the DDTs. Previous solutions for static data placed each data structure independently, but in dynamic applications multiple (actually, an unknown number of) instances of a single DDT may be created and destroyed during the execution of the application.
2. Reducing the total footprint of the groups by joining DDTs with complementary footprints. Creating a group for every DDT is close to a worst-case design because every pool later built from the groups needs enough space to accommodate the maximum number of simultaneously alive instances of the DDT. Therefore, a significant portion of the pool space may remain unused during long periods of the execution time, but these resources will not be available to create instances of DDTs from other pools.
3. Minimizing the possibility that instances of seldom accessed DDTs preclude the placement of instances of more accessed DDTs in the better memory resources. To this end, the grouping algorithm joins several DDTs in a group only if their

FPBs are very similar (the instances of both DDTs will benefit equally from being assigned to the same resources) or if their footprints are complementary, that is, if the instances of the least accessed DDTs are alive only during periods of low footprint of the most accessed DDTs (put another way, their instances will typically not compete for memory space).

7.2.2 Liveness and Exploitation Ratio

We use the following two metrics during the grouping process to minimize the periods when memory resources are underused:

- **Liveness** As calculated during the profiling step. By looking into the liveness graph for each data type, it is possible to identify variations in its instantaneous footprint such as high peaks, plateaus and sawteeth [58].
- **Exploitation ratio** The occupation degree of a pool along several time instants:

$$Exploitation\ ratio = \frac{\sum_{t=1}^{N} \frac{Required\ footprint(t)}{Pool\ size}}{N} \qquad (7.1)$$

While we compute the exploitation ratio exclusively for groups, liveness and FPB can be applied both to DDTs and groups. The liveness of a group is the evolution along time of the combined footprint of all the DDTs that it contains. The grouping algorithm tries to identify the DDTs whose liveness is complementary along time to reduce their combined footprint. It uses the liveness of a group to calculate the exploitation ratio of the future pool, and that is in turn used to decide whether joining or keeping apart the DDTs. In order to achieve a better use of resources, the algorithm may also add DDTs with a lower FPB to a group if the exploitation ratio is increased and the total size is not augmented.

We can illustrate these concepts and the aforesaid trade-off using a hypothetical application with two threads as an example. The first thread processes input events as they are received, using DDT_1 as a buffer. The second one consumes the instances of DDT_1 and builds an internal representation in DDT_2, reducing the footprint of DDT_1. Accordingly, the footprint of DDT_2 is reduced when the events expire and the related instances are deleted. Figure 7.4a shows the liveness of each DDT. The maximum footprints of the DDTs are 7 and 10 KB, respectively. Therefore, if both DDTs were treated individually, the total required footprint would be 17 KB (labeled as "Added" in the figure). However, grouping both DDTs together reduces the maximum footprint of the combined group to 11 KB (labeled as "Combined"). Figure 7.4b shows the exploitation ratio of the combined group, compared with the exploitation ratio that would result if the two DDTs were kept apart. In this hypothetical case, grouping reduces the required footprint and increases the exploitation ratio of the memory resources. This two DDTs can be kept apart from the rest of DDTs of the application, or they might be further combined with others.

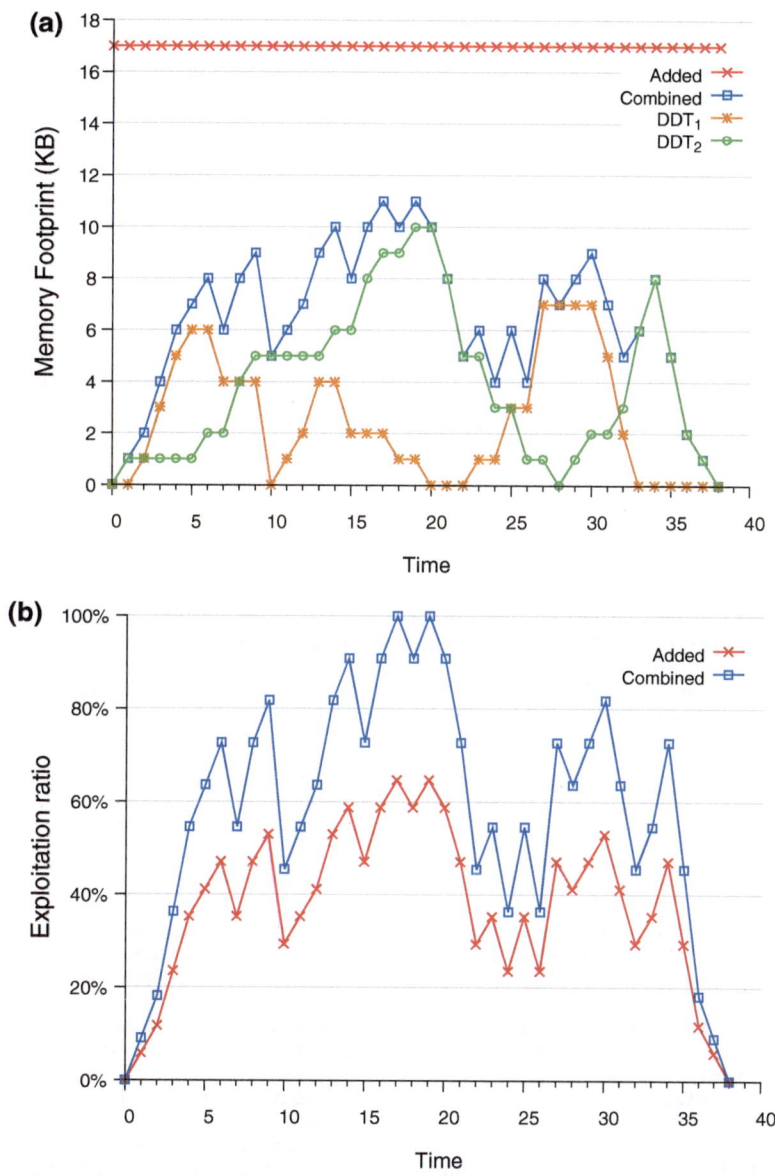

Fig. 7.4 Liveness and exploitation ratio for two hypothetical DDTs. **a** Liveness (evolution of footprint) for the two DDTs considered independently, their added maximum footprints and the footprint of a group combining them. **b** Exploitation ratio for the worst case (38.2 % on average) and the combined group (59.1 % on average)

7.2.3 The Algorithm

Algorithm 7.2 Grouping

```
 1: function Grouping(DDTs : List of DDTs) : List of Groups
 2:     Order the DDTs on descending FPB
 3:     Exclude the DDTs that were marked by the designer (ExcludedDDTs)
 4:     While there are any DDTs remaining and MaxGroups is not reached do
 5:         Create a new group
 6:         For each remaining DDT do
 7:             Calculate the liveness and FPB that would result if the DDT
         were added to the group (CALCNEWFOOTPRINTANDFPB)
 8:                 If the new behavior passes the tests in CHECKCOMPATIBILITY
         then
 9:                     Add the DDT to the group
10:                     Remove the DDT from the list of DDTs
11:     Push any remaining DDTs into the last group
12:     Add the DDTs that were excluded to the last group
13:     Order the groups on descending FPB
14:     Return the list of groups
15: end function

16: function CheckCompatibility(newBehavior : Behavior) : Boolean
17:     Return true if the new footprint does not exceed the maximum
         footprint for any group
18:         AND the footprint is not incremented more than MaxIncMF_G
19:         AND the FPB of the group is increased by at least MinIncFPB
20:         AND the exploitation ratio is increased by at least
         MinIncExpRatio
21: end function

22: function CalcNewFootprintAndFPB(Group, DDT) : Behavior
23:     Create a new behavior
24:     While there are events left in the behavior of the group or the DDT
         do
25:         Select the next event from the group and the DDT
26:         If both correspond to the same time instant then
27:             Create a new event with the sum of the footprint of the
         group and the DDT
28:         Else if the event of the group comes earlier then
29:             Create a new event with the addition of the current
         footprint of the group and the last known footprint of the DDT
30:         Else                      ▷ ( if the event of the DDT comes earlier )
31:             Create a new event with the addition of the current
         footprint of the DDT and the last known footprint of the group
32:     Update the FPB with the maximum footprint registered and the sum of
         reads and writes to the group and the DDT
33: end function
```

Algorithm 7.2 presents the pseudo-code for the grouping process. The grouping step begins with the identification of the DDTs that have the highest FPB. Previous knowledge on static data optimization has shown that objects with a higher FPB should be placed in the most efficient memories. However, dealing with instances of dynamic data structures, which are continuously created and destroyed during the execution of the application, requires additional considerations, as explained before. If including a DDT keeps the maximum footprint within a threshold because its peak requirements match the footprint minima of the group, then the DDT is added. This increases the exploitation ratio of the group as more instances will be created using the same amount of resources. The algorithm updates the group liveness as new DDTs are added.

The DDTs are evaluated in order of decreasing FPB to ensure that if a DDT matches the exploitation ratio of a group, it is the DDT with the highest FPB among the remaining ones. Once the FPB of the new group is known, a check is made to verify that a minimum increment is achieved. This is useful for instance to avoid including DDTs with low FPBs that could hinder the access pattern of the group once it is placed into memory resources such as DRAMs.

The group liveness is kept updated as a combination of the liveness of the DDTs in it. This ensures that no comparisons between individual DDTs, but between the DDTs and the groups, are done, reducing the algorithm complexity. To evaluate the inclusion of a DDT in a group, the new combined behavior is calculated (lines 22–23). This is a straightforward process that involves combining two ordered lists and accounting for the accumulated footprint and accesses of the group and the DDT. Then, the constraints imposed by the grouping parameters are evaluated for the new behavior (lines 16–21). New criteria such as the amount of consecutive accesses or specific access patterns can be easily incorporated into this check in the future. Once a DDT is definitively included in a group, only the behavior of the group as a whole is relevant and the DDT characteristics are discarded to save memory in the optimization tool.

The output of the grouping step is a list of groups with the DDTs included in each of them and their aggregated characteristics (maximum footprint, liveness, FPB and exploitation ratio). The computational complexity of the algorithm is bounded by $O(n^2 \times m)$, where n is the number of DDTs in the application and m is the number of entries in the liveness of the DDTs.

7.2.4 Algorithm Parameters

The designer can use the following parameters to guide the grouping algorithm:

1. **MaxGroups** The maximum number of groups that the algorithm can create. This parameter controls the trade-off between memory footprint and the overhead of managing multiple pools. It should be at least as big as the number of memory modules of distinct characteristics present in the platform; otherwise, the chance to separate DDTs with different behaviors into independent memory modules could be lost.
2. **MaxIncMF$_G$** The maximum footprint increase ratio when adding a DDT to a group G.

3. **MinIncFPB** Minimum ratio between the old and the new FPBs of the group to add a DDT. The default value of 1.0 allows any DDT that increases the FPB of the group to be included.
4. **MinIncExpRatio** The minimum increase in the exploitation ratio that allows a DDT with a lower FPB to be included in a group. The default value of 1.0 allows any DDT that increases the exploitation ratio of the group to be included in it. In combination with the two previous parameters, balances between increasing the exploitation ratio of the pools, maximizing the FPB of their DDTs and bounding the number of distinct DDTs that can be added to a group.
5. **ExcludedDDTs** If the designer has good knowledge of the application, it is possible to take the decision of excluding a DDT from the grouping process and manually placing it in the last group. This can also reduce the execution time of the optimization tool.

7.3 Experimental Results

We have prepared a set of experiments composed of application models and synthetic benchmarks to explore the effectiveness of our methodology with different types of applications. In essence, these experiments compare the cost of executing an application in a platform with hardware-based caches or in a platform with one (or several) explicitly addressable on-chip SRAMs managed with the methodology presented in this chapter. The first case models a network of wireless sensors where the devices have to process a moderate amount of data; however, they have to reduce their energy consumption as much as possible to increase their battery life. In contrast, the second example is a case of high performance data processing that uses the core of a network routing application. Finally, the third one is representative of DDT-intensive applications.

For each of the case studies we have executed all the steps in the methodology up to the simulation, using the following parameter values for the grouping and mapping algorithms:

- $MaxGroups = +\infty$;
- $MinIncFPB = 1.0$ and $MinIncExpRatio = 1.0$: Any increase on FPB or exploitation ratio is accepted;
- $SpreadPoolsInDRAMBanks = True$;
- $MinMemoryLeftOver = 8\ Bytes$;
- $PoolSizeIncreaseFactor = 1.0$;
- $PoolSizeIncreaseFactorForSplitPools = 1.3$;
- $UseBackupPool = True$.

With these values as a common base line for the experiments, our methodology is already able to improve on the solutions obtained with existing techniques. Further improvements might be achieved if the designers have time to fine tune them to their particular applications.

Table 7.1 Technical parameters of the cache memories

Size	Associativity	Line size words	Energy (nJ)	Latency cycles	Area (mm²)
4 KB	Direct	16	0.154499	1	0.021
32 KB	2 ways	16	0.101678	1	0.075
64 KB	4 ways	16	0.119260	1	0.140
16 KB	16 ways	16	0.166158	1	0.137
32 KB	16 ways	16	0.166158	1	0.203
32 KB	16 ways	4	0.023916	1	0.100
64 KB	16 ways	16	0.179300	1	0.263
64 KB	16 ways	4	0.030176	1	0.158
256 KB	16 ways	16	0.250323	2	0.609
256 KB	16 ways	4	0.068477	2	0.533
512 KB	16 ways	16	0.344979	2	1.069
1 MB	16 ways	16	0.509083	4	2.088
4 MB	16 ways	16	2.124347	6	8.642

7.3.1 Description of Memory Organizations

We have used a big number of memory organizations to test our methodology with the subject applications. Here, we present results only for the most relevant cases. For each of the experiments we have chosen the most relevant configurations; hence the slight differences between the platforms shown for each experiment. The technical parameters of the SRAM and cache memories were obtained via Cacti [76] using a 32 nm feature size. Tables 7.1 and 7.2 detail their respective technical parameters. The energy values shown in both tables include the cost of the address decoders and data multiplexers. The width of each access is one word (32 bits) for SRAMs and one full line for caches.

The cache memories were configured using as reference the ARM Cortex-A15 [9]: 64-byte (16 words) line size, associativity of up to 16 ways, up to 4 MB in size (for the L2 caches) and Least Recently Used (LRU) replacement policy. We present many different configuration cases to explore the implications of these design options. Among others, we used configurations with sizes from 4 KB up to 4 MB; direct mapped ("D"), 2-ways ("A2"), 4-ways ("A4") and 16-ways ("A16") associativity; 16 (default) and 4 ("W4") words per line (Fig. 7.5). All cache memories use an LRU replacement policy. Finally, we include configurations labeled as *"lower bound"* that represent the minimum theoretical cost for a cache memory with a big size (256 MB), but a small cost (comparable to that of a 4 KB one).

The configurations managed with our methodology comprise one or several on-chip SRAMs (Fig. 7.6) and a external DRAM. These configurations are labeled as "SRAM", where their labels enumerate the independent memory modules that are included. For instance, a configuration labeled as "SRAM 8 × 512 KB" has 8

Table 7.2 Technical parameters of the (on-chip) SRAMs

Size	Energy (nJ)	Latency cycles	Area (mm²)
512	0.000648	1	0.003
1 KB	0.001096	1	0.005
4 KB	0.001649	1	0.012
16 KB	0.003710	1	0.045
32 KB	0.004924	1	0.112
64 KB	0.007230	1	0.185
256 KB	0.012558	2	0.781
512 KB	0.024775	2	1.586
1 MB	0.027549	4	2.829
4 MB	0.076838	6	11.029

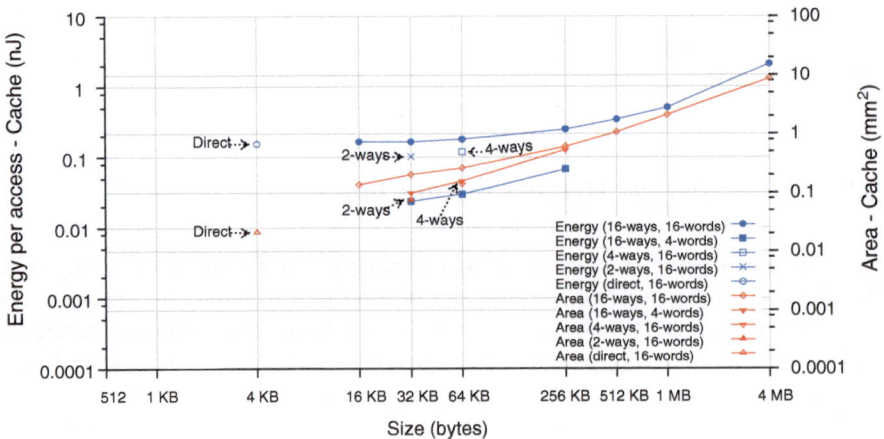

Fig. 7.5 Energy per access and area for cache memories of varying sizes. All axes are in logarithmic scale

modules of 512 KB each. Through all the tables in the experiments, we include a base configuration with SRAM modules of 512 B, 1, 32 and 256 KB as a common reference. Finally, we also include a configuration labeled as "lower bound" that represents the minimum theoretical costs for an on-chip SRAM of 1 GB with the properties of a 4 KB memory. As stated previously, our approach can be easily combined with the use of caches. Although in the set of experiments of this work the effect of these combinations is not relevant, they may become useful in cases where a small set of data instances gets a low but still significant number of accesses (with some locality) in the DRAM modules.

Finally, we modeled the DRAM modules according to two different technologies designed for embedded systems: low power (mobile) SDRAM [119] and LPDDR2-

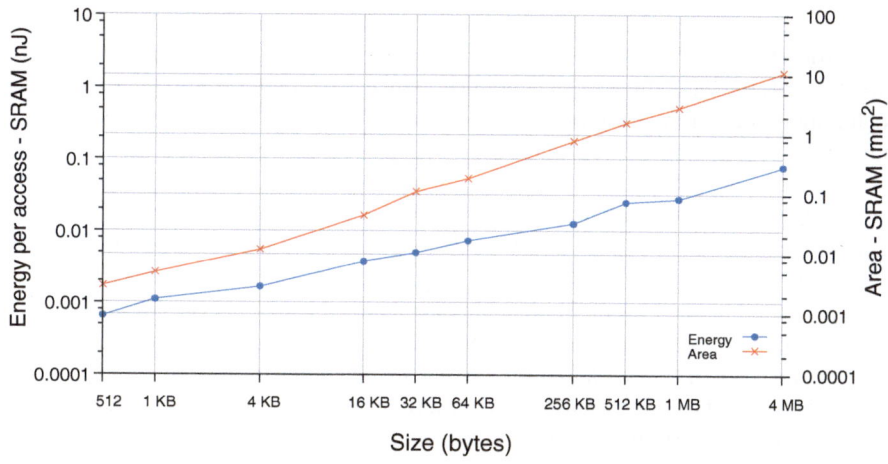

Fig. 7.6 Energy per access and area for SRAM memories of varying sizes. All axes are in logarithmic scale

Table 7.3 Technical parameters of the Mobile SDRAM and LPDDR2-SDRAM memories

Type	Size (MB)	Other parameters (units in MHz, V, mA, pF)
Mobile	128	166 MHz, 4 banks, 1 024 words/row, CL = 3, tRP = 3, tRCD = 3, tWR = 3, tCDL = 1, tRC = 10, vDD = 1.8, vDDq = 1.8, iDD0 = 67.6, iDD1 = 90, iDD2 = 15, iDD3 = 18, iDD4 = 130, iDD5 = 100, iDD6 = 15, cLoad = 20
LPDDR2	256	333 MHz, 8 banks, 2 048 words/row, tRL = 5, tWL = 2, tRCD = 6, tDQSCK_SQ = 1, tDQSS = 1, tCCD = 1, tRTP = 3, tRAS = 14, tRPpb = 6, tWTR = 3, tWR = 5, vDD1 = 1.8, vDD2 = 1.2, vDDca = 1.2, vDDq = 1.2, iDDO1 = 20, iDDO2 = 47, iDDOin = 6, iDD3N1 = 1.2, iDD3N2 = 23, iDD3Nin = 6, iDD4R1 = 5, iDD4R2 = 200, iDD4Rin = 6, iDD4Rq = 6, iDD4W1 = 10, iDD4W2 = 175, iDD4Win = 28, cLOAD = 20

SDRAM [81]. The technical parameters for the LPSDRAM and LPDDR2-SDRAM modules, shown in Table 7.3, were obtained from the manufacturer's data sheets [119, 120].

7.3.2 Case Study 1: Wireless Sensors Network

In this first case study we apply the methodology step by step on an application, optimizing it for several platform configurations with cache or SRAMs of equivalent sizes. In this way, we show that the approach presented in this chapter can provide important optimizations. The subject application is a model of a network of wireless sensors that can be distributed over wide areas. Each sensor extracts information of its surroundings and sends it through a low-power radio link to the next sensor in the path towards the base station. The networks constructed in this way are very resilient to failures and terrain irregularities because the nodes can find new paths. They may be used for applications such as weather monitoring [79] and fire detection in wild areas or radio signal detection in military operations. Each sensor keeps several hash tables and input and output queues to provide some basic buffering capabilities. The sensors use dynamic memory due to the impossibility of determining the network dependencies and the need of adjusting the use of resources to what is strictly needed at each moment. As a consequence, the sizes of the hash tables and network queues, among others, need to be scaled continuously. The application model creates a network of sensors and monitors one in the middle of the transmission chain; that node forwards the information from more distant ones and gathers and sends its own information periodically. In the next paragraphs we explain how the different steps of the methodology are performed and their outcome.

7.3.2.1 Profiling

Adding the profiling instrumentation to the application requires modifying 9 lines out of 3 484. That means that just a 0.26 % of the source code has to be modified both for profiling and deployment (or simulation). After instrumentation, the application was executed normally during 4 hours of real time, producing a log file of 464 MB.

7.3.2.2 Analysis

Our tools analyze the log file obtained through profiling in about 29 s, identifying 21 distinct DDTs. The FPB of the DDTs range from 8.8×10^5 down to 0.61 accesses per byte. The maximum footprint required for a given DDT is 24 624 B and the minimum, 12 B. The size of the hash tables varies during the application execution; for instance, the hash table for the active neighbors uses internal arrays of 804 B, 1644 B, 3324 B, 6684 B and 13404 B. Our tools detect this and separate the different instances of the hash table DDT (classifying them according to their different sizes) for the next steps.

7.3.2.3 Grouping

The grouping step is executed (in about 70 s) with the parameters explained at the beginning of this section. Our tool builds a total of 12 groups from the initial set of 21 DDTs. One of the groups has 5 DDTs, one group gets 3 DDTs, three groups hold 2 DDTs and the last seven groups contain just 1 DDT. The grouping step manages to reduce the total footprint of the application (compared to the case of one DDT per pool) from 110 980 B to 90 868 B, thus achieving an overall footprint reduction of

an 18.12% and reducing from 21 to 12 the number of pools that have to be managed
(−42.9 %). For the five groups that the tool generates combining several DDTs, the
respective memory footprint reductions are 32.6, 56.4, 19.9, 51.7 and 19.6 % when
compared to the space that would be required if each DDT were mapped into an
independent pool.

Figure 7.7 gives more insights into the grouping step. The tool identifies DDT_2
and DDT_6 from the application as compatible and combines them in the same group,
yielding a significant reduction in memory usage. If each DDT is mapped into an
independent pool, as is done in techniques that build per-DDT pools, the total memory
space required is 1,032 B (thick black horizontal line in the graph). With grouping,
the required footprint is reduced to just 696 B (green horizontal line), a reduction
of 32.6 %.

Additionally, Fig. 7.7a shows that the maximum memory footprint for the DDTs
and the group is determined by instantaneous peaks in the graph. The designer might
use the *PoolSizeIncreaseRatio* parameter during the mapping phase to adjust the final
footprint so that it covers only the space required during most of the time, relying
on the use of a backup pool in the DRAM to absorb the peaks. However, there is
an important drawback to this. Although the freed space could then be exploited by
other DDTs (instead of staying unused when the footprint of the group is lower), it is
possible that the instances created during the peak footprint periods, and that would
have to be allocated in the backup pool, get so many accesses that the total system
performance is reduced. After all, the instances of the DDTs included in the group
have a higher FPB than the instances of DDTs included in the next groups. This
factor stresses the importance of the grouping step to improve the exploitation ratio
of the memory space assigned to each DDT and of the simulation step to predict the
consequences of the different parameters.

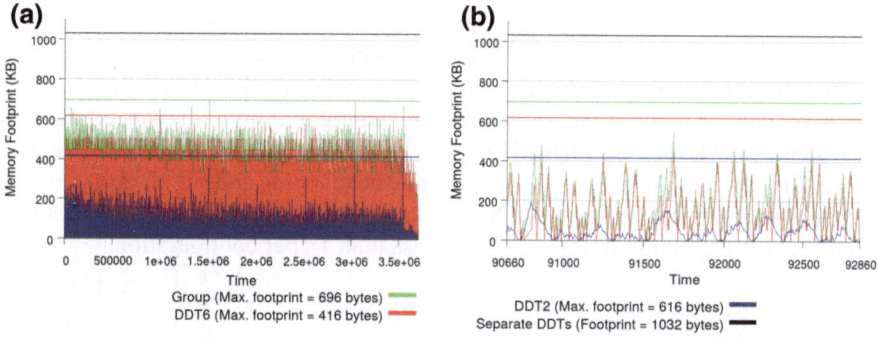

Fig. 7.7 Case study 1: analysis of the footprint of two DDTs in the application (*red* and *blue plots*)
and the group that holds both of them (*green plot*). The combined footprint of the group (*green
horizontal bar*) is significantly lower than the added footprint that would be needed if the DDTs
were assigned to independent pools (*black horizontal bar*). **a** Plots for the complete execution time
of the application. **b** Zoom over a randomly-chosen area of 2 200 (allocation-time) instants

Finally, the evolution of the footprint of the two DDTs and the combined group can be observed in more detail in Fig. 7.7b. There, we present a randomly-chosen period during the execution time of the application. The fact that the footprint peaks of DDT_2 are not coincident with the peaks of DDT_6 is the factor that allows for their grouping.

7.3.2.4   Pool Formation

For this experiment, we use an always-coalesce-and-split scheme with free lists per size. When a block is allocated, any extra space is separated and inserted into a list of free blocks; when a block is de-allocated, if the previous or next blocks are also free, they are fused and inserted into the matching list. In these experiments, the simulator does not consider the accesses of the dynamic memory managers (only the application accesses are traced). Therefore, our simulator cannot be used currently for analyzing the impact of the different DMMs generated by the techniques explained in Chap. 6. This is due to the fact that the simulator executes the traces of memory accesses obtained during profiling. It does not execute real platform code; hence it is unable to track the accesses generated by different versions of the code. Nevertheless, it should be straightforward to introduce this functionality by linking the library of DMMs in the simulator itself so that every time that the DMM code needs to read or update an internal structure, these accesses are reproduced in the simulated platform.

7.3.2.5   Mapping

The mapping step is the first one in the methodology that is platform-dependent. In this case, we run the mapping and simulation steps for 230 different platforms to verify that our methodology achieves better results than cache memories for many different potential platform configurations. Due to the small footprint of this application, we include in these experiments several configurations with reduced memory subsystems (down to just 32 KB). However, it is important to stress that, if the platform is already decided, the designer has to run the mapping and simulation steps just once with the platform parameters. Our tool can execute this step with the parameters detailed at the beginning of this section in less than 1 s per configuration.

7.3.2.6   Simulation

We analyze the performance of every mapping solution over its corresponding platform using the simulator explained in this chapter. The simulations use the original memory trace obtained during the profiling step and require about 16 s each (the simulation process is less complex than the analysis). As the DRAM modules are

Table 7.4 Case study 1: Performance of the solutions obtained with our methodology versus cache-based solutions

Platform	Energy mJ	Energy %	Time ($\times 10^6$) cycles	Time %	Page misses %	DRAM accesses %	Total accesses %
(Mobile SDRAM)							
00. (only DRAM)	360.01	100.0	2132.5	100.0	100.0	100.0	100.0
01. Cache: L1 = 256KB(A16)	21.88	6.1	174.5	8.2	<0.1	<0.1	100.1
02. Cache: L1 = 256KB(A16,W4)	6.03	1.7	174.5	8.2	<0.1	<0.1	100.1
03. Cache: L1 = 16KB(A16), L2 = 256KB(A16)	52.21	14.5	270.2	12.7	0.1	0.3	301.0
04. Cache: L1 = 32KB(A16), L2 = 256KB(A16)	34.24	9.5	182.4	8.6	0.1	0.4	202.7
05. SRAM:512 B, 1, 32, 256 KB	**0.24**	**0.1**	**90.1**	**4.2**	**0.0**	**0.0**	**100.0**
06. Cache: L1 = 64KB(A16)	15.94	4.4	88.3	4.1	<0.1	0.1	100.2
07. Cache: L1 = 64KB(A16,W4)	2.94	0.8	88.8	4.2	0.1	0.1	100.2
08. SRAM:512 B, 1, 64 KB	0.46	0.1	87.7	4.1	<0.1	0.1	100.0
09. SRAM:64 KB	0.93	0.3	88.1	4.1	<0.1	0.1	100.0
10. Cache: L1 = 32KB(A16)	106.71	29.6	434.3	20.4	9.7	35.0	168.7
11. Cache: L1 = 32KB(A16,W4)	36.08	10.0	275.4	12.9	11.5	9.0	118.0
12. SRAM:512 B, 1, 32 KB	10.62	2.9	132.9	6.2	2.2	3.4	100.0
13. SRAM:512 B, 1, 16, 32 KB	6.39	1.8	112.3	5.3	1.0	2.1	100.0
14. SRAM:32 KB	12.23	3.4	141.1	6.6	2.9	3.7	100.0
15. SRAM:LowerBound	0.14	<0.1	87.1	4.1	0.0	0.0	100.0
16. Cache: LowerBound(D)	13.52	3.8	87.3	4.1	<0.1	<0.1	100.1
(LPDDR2-SDRAM)							
17. (only DRAM)	236.56	100.0	1315.6	100.0	100.0	100.0	100.0
18. Cache: L1 = 256KB(A16)	21.84	9.2	174.4	13.3	<0.1	<0.1	100.1
19. Cache: L1 = 256KB(A16,W4)	6.01	2.5	174.5	13.3	<0.1	<0.1	100.1
20. Cache: L1 = 16KB(A16), L2 = 256KB(A16)	55.10	23.3	285.1	21.7	0.3	0.5	318.8
21. Cache: L1 = 32KB(A16), L2 = 256KB(A16)	33.29	14.1	178.8	13.6	0.3	0.5	200.5
22. SRAM:512 B, 1, 32, 256 KB	**0.24**	**0.1**	**90.1**	**6.8**	**0.0**	**0.0**	**100.0**
23. Cache: L1 = 64KB(A16)	16.25	6.9	90.0	6.8	0.5	0.6	101.1
24. Cache: L1 = 64KB(A16,W4)	2.90	1.2	89.1	6.8	0.4	0.1	100.2
25. SRAM:512 B, 1, 64 KB	0.40	0.2	87.4	6.6	0.1	0.1	100.0
26. SRAM:64 KB	0.86	0.4	88.0	6.7	0.1	0.1	100.0
27. Cache: L1 = 32KB(A16)	50.64	21.4	258.5	19.6	26.9	34.3	167.3
28. Cache: L1 = 32KB(A16,W4)	35.84	15.2	323.6	24.6	47.6	15.8	129.7
29. SRAM:512 B, 1, 32 KB	7.88	3.3	117.6	8.9	0.3	3.4	100.0

(continued)

Table 7.4 (continued)

30. SRAM:512 B, 1, 16, 32 KB	4.99	2.1	106.0	8.1	0.1	2.1	100.0
31. SRAM:32 KB	9.44	4.0	126.8	9.6	3.2	3.7	100.0
32. SRAM:LowerBound	0.14	0.1	87.1	6.6	0.0	0.0	100.0
33. Cache: LowerBound(D)	13.48	5.7	87.2	6.6	<0.1	<0.1	100.1

The entries in the first half of the table, which correspond to platforms with a Mobile SDRAM, are normalized using as reference platform 00. Correspondingly, the entries in the second part (platforms with an LPDDR2), are normalized taking as reference platform 17

seldom accessed in most platforms, we assume that the memory controller can drive the chips into one of the power-saving modes and thus we discard their active-idle energy consumption. We take as reference the platforms with only DRAM modules (that is, without cache memories or SRAMs), which correspond to platforms 00 and 17 in this experiment; the normalized results are presented in Table 7.4.

The results of the experiments show that even for small sizes, cache memories improve significantly the performance of the system. However, the solutions obtained with the methodology presented in this chapter achieve even bigger gains. To get a better measure of this improvement, we show in Fig. 7.8 a direct comparison between both types of solutions for various memory subsystem sizes. These results are normalized, for each size, to the performance of the corresponding cache (the leftmost bars at the left of each group). For instance, platform 06 has a 64 KB cache memory and reduces the energy consumption in comparison with the reference case, platform 00, down to a 4.4 %. In the figure, this is taken as the base case for the group of bars in the middle. Modifying the length of the cache lines (platform 07) reduces the energy consumption down to an 18.4 %. However, using instead an SRAM of the same size (platform 09), energy consumption is reduced down to 5.9 % of the energy consumption with the 64 KB cache. Furthermore, this still represents a reduction in energy consumption of a 68.3 % from platform 07 (the one with shorter cache lines).

The graphs show a different evolution for the number of cycles than for the energy consumption because latencies do not scale linearly with the size of the caches. For instance, we used a latency of one clock cycle for all the memories smaller than 128 KB, whether caches or SRAMs. Also, we estimated that SRAMs and caches of the same size have the same latency.

7.3.3 Case Study 2: Network Routing

In this case study we explore the application of the methodology to a multi-threaded implementation of the core algorithms in the network dispatcher of an operating system. The model is similar to the one presented in the case study of Chap. 3. Embedded systems may execute threads from several applications concurrently; the operating system has then to arbitrate between the threads that need to send data through the network. As explained there, a common choice in these cases is the Deficit

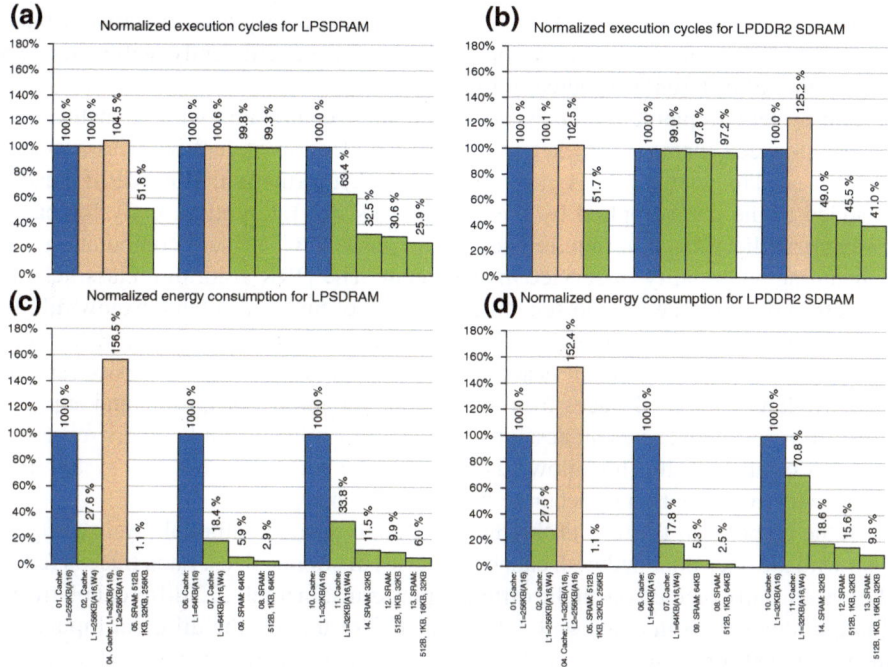

Fig. 7.8 Case study 1: Comparison of our SRAM-based solutions with solutions based on caches of equivalent capacity. Each group of results is normalized to the leftmost bar (*in blue*), which represents the performance for a cache of a given size

Round Robin (DRR) algorithm [148]. The operating system keeps a list of active destinations and, when a thread sends a packet towards one of them, it is stored in the corresponding queue. Packets are extracted from the queues in order and forwarded to the network adaptor, reducing the credit of the queue proportionally to the size of the packet. This mechanism allows for the implementation of Quality-of-Service (QoS) mechanisms that can prioritize among applications and destinations, for instance, to guarantee a minimum bandwidth for certain connections while avoiding starvation in the rest. The implementation that we have used is multi-threaded. Therefore, several instances of the application DDTs are alive and accessed at the same time by different threads; moreover, memory accesses do not happen in a sequential manner. The main issue here is that it is not possible (at least in an easy way) to analyze the memory access behavior of each thread independently from each other and then, by combining these individual behaviors, "recreate" the behavior of the whole system, because that would leave out the interaction among the threads and, most importantly, the interleaved thread execution. Consider, for example, the case of two threads where each of them generates a stream of sequential accesses to memory: The memory subsystem would receive the accesses interleaved. Indeed, if the system has multiple processors or SMT (simultaneous multi-threading) capabilities, the accesses will be interleaved every few words. If the involved DDTs have been placed on the same bank of a DRAM, these streams, that would otherwise execute efficiently, will generate a big number of different row activations.

The system implementation is organized in five modules with asynchronous queues between them: (a) Packet injection: A collection of real wireless network traces is used to generate the packets sent; (b) Packet formation: The TCP/IP header is added to the data supplied by the applications; (c) Encryption: Only for packets from applications that require encryption; (d) TCP Checksum computation; (e) Scheduling and quality-of-service management: The DRR algorithm classifies the packets in priority queues and schedules them according to available bandwidth. In order to reproduce the conditions of a real system, we have used a set of network traces collected from the wireless access points of the Dartmouth University campus [73]. We have identified traces from individual, yet anonymous, users and applications; some of them represent sessions lasting a few minutes while others represent sessions of up to 24 h. This allows us to reproduce part of the original system use cases, with the exception of accurate packet rate control.

Table 7.5 presents the results obtained after applying our methodology, in comparison with the ones obtained with cache memories of similar sizes. Each trace is fed to the system and executed on all the different platforms. In total, 222 different platform configurations are simulated with 14 traces. The results for the different inputs are then averaged for each platform. The sample standard deviation in the table shows important fluctuations from the average values for almost every platform because of the different nature and length of the inputs. However, the solutions generated with the methodology of this chapter improve in all cases on the results obtained with caches. Interestingly, the standard deviation of our solutions is smaller than that of cache solutions. This suggests that our methodology tackles better with the dynamic conditions of the application and achieves a more uniform system performance.

As a final consideration, in Table 7.5 energy costs are normalized independently for the platforms with Mobile SDRAM and LPDDR2. However, a direct comparison between them exposes a net reduction of a 41 % on average for all platforms (s = 14.7 %) when using an LPDDR2 instead of the older technology (specifically, 70.9 % with s = 10.3 % when comparing platforms 21 and 7, excluding the experiments that fit entirely in the on-chip memories). This highlights the important effort invested by the industry in reducing the energy consumption of the memories designed for embedded systems. Modern DRAM technologies such as LPDDR3 and LPDDR4 (introduced in 2014) should be able to reduce even further the cost of accessing the main DRAM.

In this experiment, footprint varies approximately from 242 KB to 8.7 MB, with 13 out of the 14 cases over 512 KB. This situation attests that the solutions produced by this methodology have good performance also when the application footprint exceeds the size of the available on-chip SRAMs, even if it avoids data movements across memory elements. The reason for this good performance is that the data placement puts in the DRAM only instances of the least accessed DDTs and, in this case, those instances (the body of the network packets) are accessed mostly sequentially. Temporal locality is very low because each instance is accessed just twice, first to

Table 7.5   Case study 2: performance of the solutions obtained with our methodology versus cache-based solutions

Platform	Energy	σ_{n-1}	Time	σ_{n-1}	Page misses	σ_{n-1}	DRAM accesses	Total accesses
(Mobile SDRAM)								
00. (only DRAM)	100.0	0.0	100.0	0.0	100.0	0.0	100.0	100.0
01. Cache: L1 = 256KB(A16)	46.6	30.3	39.5	26.1	7.3	3.5	43.2	184.1
02. Cache: L1 = 256KB(A16,W4)	38.9	25.6	40.9	23.5	16.5	7.0	41.3	173.2
03. Cache: L1 = 512KB(A16)	42.8	37.1	34.4	29.6	4.3	4.8	33.5	165.0
04. Cache: L1 = 4MB(A16)	107.0	66.5	54.7	36.2	3.1	4.3	28.6	155.5
05. Cache: L1 = 16KB(A16), L2 = 256KB(A16)	52.3	33.1	41.8	28.9	7.5	3.0	44.8	294.8
06. Cache: L1 = 32KB(A2), L2 = 256KB(A16)	49.6	32.3	41.8	28.9	7.4	3.0	44.8	295.9
07. SRAM: 512 B, 1, 32, 256 KB	**25.0**	**25.8**	**22.7**	**19.6**	**0.8**	**0.8**	**31.4**	**100.0**
08. SRAM: 256 KB	25.9	25.6	25.8	18.8	0.8	0.9	32.0	100.0
09. SRAM: 4 MB	15.3	14.3	36.5	12.9	0.3	0.4	12.6	100.0
10. SRAM: 8 × 512 KB	11.3	14.9	17.6	11.8	0.3	0.4	12.6	100.0
11. SRAM: 512 B, 1, 32, 256, 8 × 512 KB	10.1	13.9	14.2	11.9	0.3	0.4	11.7	100.0
12. SRAM: LowerBound	<0.1	0.0	5.7	1.7	0.0	0.0	0.0	100.0
13. Cache: LowerBound(D)	6.9	3.2	7.0	2.5	<0.1	0.1	2.7	105.2
(LPDDR2-SDRAM)								
14. (only DRAM)	100.0	0.0	100.0	0.0	100.0	0.0	100.0	100.0
15. Cache: L1 = 256KB(A16)	46.3	35.0	34.9	24.5	5.2	2.8	43.2	184.1
16. Cache: L1 = 256KB(A16,W4)	55.7	49.4	56.9	44.5	10.1	5.1	41.3	173.2
17. Cache: L1 = 512KB(A16)	49.0	43.1	32.0	26.4	2.5	3.0	33.5	165.0
18. Cache: L1 = 4MB(A16)	184.8	140.1	64.0	42.5	1.4	2.0	28.6	155.5
19. Cache: L1 = 16KB(A16), L2 = 256KB(A16)	57.3	45.3	38.4	29.9	5.9	2.1	44.8	294.7
20. Cache: L1 = 32KB(A2), L2 = 256KB(A16)	51.8	42.0	38.4	30.0	5.9	2.1	44.8	295.9
21. SRAM: 512B, 1, 32, 256 KB	**11.1**	**11.6**	**12.9**	**8.1**	**0.9**	**0.9**	**31.4**	**100.0**
22. SRAM: 256 KB	12.1	11.6	16.7	7.9	0.9	0.9	32.0	100.0
23. SRAM: 4MB	14.7	9.5	43.7	15.3	0.3	0.4	12.6	100.0
24. SRAM: 8 × 512 KB	7.0	8.5	17.0	8.4	0.3	0.4	12.6	100.0
25. SRAM: 512 B, 1, 32, 256, 8 × 512 KB	5.9	7.9	13.2	8.7	0.3	0.4	11.7	100.0
26. SRAM: LowerBound	0.1	0.0	8.4	4.1	0.0	0.0	0.0	100.0
27. Cache: LowerBound(D)	10.6	4.6	9.1	4.2	<0.1	0.0	2.7	105.3

Average normalized improvements with sample standard deviations (σ_{n-1}, sample size $N = 14$). All magnitudes are in percentages. The entries in the first half of the table, which correspond to platforms with a Mobile SDRAM, are normalized using as reference platform 00. The entries in the second part (platforms with an LPDDR2), are normalized taking as reference platform 14 calculate the CRC and then to forward it to the buffers of the network adaptor. Cache memories can save the second access to DRAM, but only if it happens soon enough to avoid being evicted by other data accessed from other threads. However, it is possible that this is done at the cost of evicting more useful cache lines. An alternative in platforms based on caches could be mapping these DDTs in a non-cacheable area, but that would deactivate caching for all the packet bodies.[3]

In contrast, as our solutions can split the pools over different memory resources, a fraction of the instances of the packet bodies are still placed on the on-chip SRAMs, but we have the security that no instances of other more accessed DDTs are evicted. If the application has a memory allocation pattern that alternates peaks with periods of lower consumption then, during a potentially significant fraction of the execution time, all the instances of all the DDTs may reside in the on-chip memories without conflicts (that is, if the footprint of the packet bodies is small enough to fit in the part of their pool mapped on the closer memories). As the number of packet bodies increases, some of them are allocated in the DRAMs, but accessing them does not evict more accessed data from other pools.

In a different consideration, the observed increase in energy consumption related to the bigger cache memories is due to the fact that bigger caches consume more energy per access, and the smaller ones are already capable of the most significant reduction of accesses to the DRAM. This effect is less clear in the number of cycles required for execution because the cache latencies used in our experiments do not increase linearly with their size. Moreover, cache hierarchies perform quite badly in this experiment. Consider for example the case of platforms 01 and 05. Despite having more capacity, the second platform has a considerably higher energy consumption. This is due to the continuous transfers of data with low locality between the small L1 and the L2. As a result, the total number of memory accesses, that is, the addition of accesses to all the memory modules, including transfers between levels in the cache hierarchy, is much higher in the second platform. Compare these results with the ones obtained for platform 08, which has an SRAM of the same size than the cache of platform 01.

Finally, we present in Table 7.6 a comparison of non-aggregated data for several input cases to compensate for the big standard deviation in the aggregated results. In that table, we compare the results for our solutions to the results of a solution with a cache of equivalent size.

---

[3] It could still be possible to split the pool of packet bodies in two areas, one cacheable and the other non-cacheable, allocating space from the first one as long as possible. However, how big should the cacheable pool be? The answer to this question would require an analysis very similar to the one we propose! Besides, due to the difficulties in predicting the run-time behavior of cache hierarchies, some unwanted interactions could still happen.

7.3.4 Case Study 3: Synthetic Benchmark

The goal of this last benchmark is to show that DM-intensive applications can limit the effectiveness of cache memories because of their low spatial and temporal localities. The benchmark uses a trie to create an ordered dictionary of English words, and then simulates multiple user look-up operations. The trie DDT [61] belongs to the category of ordered trees and is useful to store any type of information that can be organized using prefixes, especially if it presents a high degree of redundancy, such as words of a dictionary, compression tables, DNA sequences, etc. In this benchmark each node has a list of children indexed by letters.

This experiment models a case that is particularly hostile to cache memories because each traversal accesses a single word on each level, the pointer to the next

Table 7.6 Case study 2: Detailed comparison between SRAM and cache-based solutions

Platform	(Mobile SDRAM)				(LPDDR2-SDRAM)			
	Energy	Time	DRAM accesses	Page misses	Energy	Time	DRAM accesses	Page misses
Cache: L1=256KB(A16)	100.0	100.0	100.0	100.0	100.0	100.0	100.0	100.0
Cache: L1=64KB(A4)	98.9	92.9	123.8	130.2	81.2	80.4	123.9	141.1
SRAM: 64 KB	52.4	56.7	79.5	30.6	45.3	61.1	79.6	43.3
SRAM: 256 KB	24.3	53.2	32.5	2.5	15.6	61.4	32.6	3.1
SRAM: 512 B, 1, 32, 256 KB	19.7	35.5	28.7	1.9	10.4	37.5	28.7	2.2
Cache: L1=256KB(A16)	100.0	100.0	100.0	100.0	100.0	100.0	100.0	100.0
Cache: L1=64KB(A4)	94.1	93.5	104.8	131.1	80.0	83.1	104.8	134.7
SRAM: 64 KB	72.3	68.8	95.9	40.1	35.9	41.8	95.9	71.9
SRAM: 256 KB	68.3	68.2	90.7	14.5	30.1	41.4	90.7	20.1
SRAM: 512 B, 1, 32, 256	67.6	65.0	90.2	14.4	29.3	36.1	90.2	20.0
Cache: L1=256KB(A16)	100.0	100.0	100.0	100.0	100.0	100.0	100.0	100.0
Cache: L1=64KB(A4)	86.5	83.0	103.9	102.9	72.6	72.9	103.9	102.3
SRAM: 64 KB	46.8	50.9	69.5	20.2	41.7	56.7	69.5	29.7
SRAM: 256 KB	22.2	49.1	27.8	5.2	16.4	59.3	27.8	7.7
SRAM: 512 B, 1, 32, 256 KB	17.8	32.8	24.2	4.7	11.5	37.1	24.2	6.9
Cache: L1=256KB(A16)	100.0	100.0	100.0	100.0	100.0	100.0	100.0	100.0
Cache: L1=64KB(A4)	92.4	91.9	102.8	122.4	78.5	81.5	102.8	119.9
SRAM: 64 KB	73.2	69.5	96.8	38.6	34.9	40.3	96.8	78.5
SRAM: 256 KB	70.3	69.6	92.8	20.3	31.0	41.3	92.8	34.6
SRAM: 512 B, 1, 32, 256 KB	69.7	66.8	92.5	20.2	29.9	36.1	92.5	34.4

The execution cost of our solutions is normalized for each input against the cost of a solution with a cache memory of equivalent size. These results correspond to 4 of the 14 inputs used in the experiments, not aggregated. All numbers are percentages child, but the cache has to move whole lines after every conflict. This is a well known side effect of the use of dynamically-linked data structures on cache architectures, including desktop and server computers, and is thus an area of intense research.

Figure 7.9 shows the improvements attained in comparison with cache memories. Each of the subfigures shows three blocks of results. The bars in the blocks are normalized to the first bar in that block (depicted in blue), which represents the performance of the reference cache platform for that block. The first block of bars compares the performance of caches and SRAMs of increasing sizes to that of a 256 KB cache. Our solution achieves a relative improvement over the cache solution of 83.2 % for the LPSDRAM case and 77.2 % for the LPDDR2 case in energy consumption, 70.8 and 60.0 % when considering the cycles spent accessing the memories (platform "SRAM: 512 B, 1, 32, 256" versus "Cache: L1 = 256KB(A16)"). As discussed in the previous case study, cache hierarchies (platform "Cache: L1 = 32KB(A16), L2 = 256KB(A16)") may perform poorly with this type of applications because the small size of the caches in the closer levels forces many evictions of whole lines, and the transfers between cache levels accumulate to the accesses to the DRAM themselves.

One of the most interesting results shown in the figure is obtained by reducing the line-size of the cache memory.Our base case is the configuration offered by

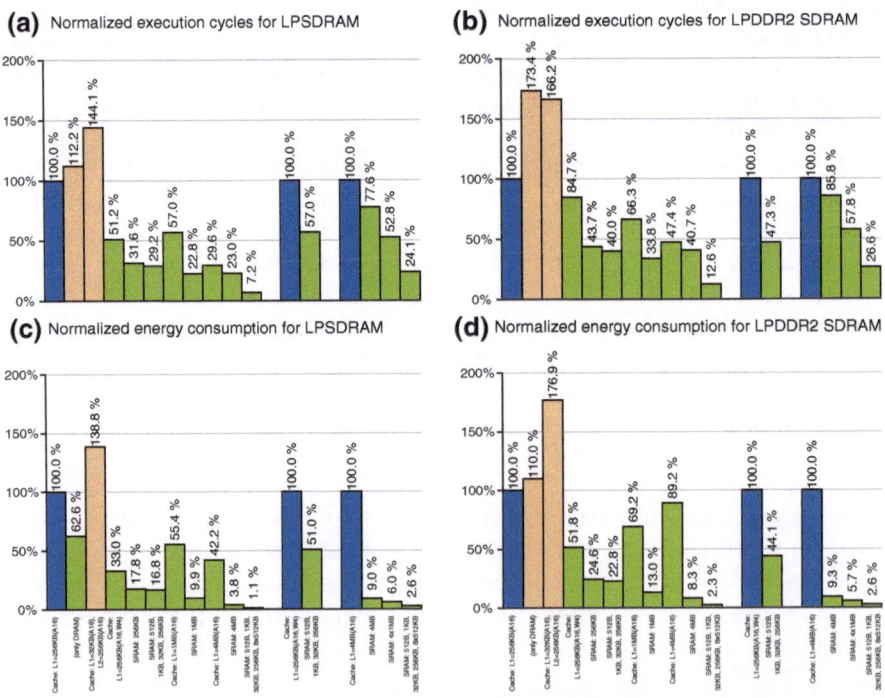

Fig. 7.9 Case study 3: comparison of SRAM-based solutions obtained with the methodology of this chapter with solutions based on caches of equivalent capacity. Each block of results is normalized to the leftmost bar (*in blue*) in that block

current ARM Cortex-A15 processors, which has 64-byte cache lines. Although the trend for larger line sizes may be beneficial for applications that process streams of data (because it increases the length of prefetching), for applications that rely heavily on the use of DDTs the effect is quite different: Every line fill and write-back is more costly because more words are moved between memories even if the application accesses just one or two of them, and the longer lines reduce the number of different cache lines (i.e., different memory positions) that can be stored with the same cache size. Our experiments show that using a line size of just 16 bytes (4 words) can improve, for this type of applications, the energy consumption by 67.0 % for the LPSDRAM and 48.2 % for the LPDDR2, and the cycles spent by 48.8 and 15.3 %, respectively (platform "Cache: L1 = 256KB(A16,W4)" versus "Cache: L1 = 256KB(A16)"). Nevertheless, our approach with explicitly addressable memories is even better suited for this type of applications, still improving energy consumption over the 16-byte-lines cache memory by 49.0 and 55.9 %, and the cycles used by 43.0 and 52.7 %, respectively (platform "SRAM: 512 B, 1, 32, 256 KB" versus "Cache: L1 = 256KB(A16,W4)", second block in all the graphs).

Finally, the performance of our solutions also scales better when the size of the memories is increased (shown in the third block of bars in all the graphs of the figure). Moreover, this last block of bars shows how multiple on-chip SRAMs of a smaller size can be combined with our techniques to further improve performance, in contrast with the difficulties encountered while trying to harness cache hierarchies.

7.4 Comparison to Related Work

In the following paragraphs we retrospectively compare the main characteristics of the methodology presented in this chapter with other previously existing techniques.

7.4.1 Static Data

A considerable amount of research effort [86, 87, 155, 167, 168] has been devoted to assign data types into heterogeneous memory hierarchies, so that the most frequently accessed data reside in the closest memories to the processor. Good overviews of several techniques to improve memory footprint and decrease energy consumption in statically allocated data (stack and global variables) are presented by [133] and [25]. The concept of selfishness to reduce the number of page misses in the banks of each DRAM module is introduced in [109]. A hardware structure to transparently map stack data into a scratchpad memory is presented by [70]. In [151], the authors explore, both from the perspective of an exact solver and using heuristics, the problem of placing (static) data structures in a memory subsystem where several memories can be accessed in parallel. Their work is partially a generalization of the mapping phase in our methodology, if each of our pools is considered as a big static data structure (an array) with a fixed size. However, their approach prevents splitting a pool over several memory resources. In this regard, it is interesting to note that a data structure is viewed in most works as an atomic entity; however, in this chapter we deal with entire pools (in contrast with individual DDTs) and consider splitting them over memory modules even if their address ranges are not consecutive. This consideration transforms a (hard) integer knapsack problem into a fractional knapsack one, which is much easier to solve and, when applicable, usually yields better solutions.

7.4.2 Software-Controlled Memories

For years, researchers have tried to improve the performance of embedded systems using software-controlled memories instead of hardware-controlled caches [67], using techniques such as tiling [4], array blocking and caching [102] to take advantage of the temporal and spatial localities of the applications, but at the expense of making continual data movements between memories. The fundamental idea was to leverage design time knowledge about the application and the presence of a Direct Memory Access (DMA) controller to produce a dynamic data layout [2, 50], scheduling copies between main memory and the scratchpad at the appropriate times. However, these techniques are only suitable if the algorithms make repeated accesses to the data that is brought closer [176]. Otherwise, performance may still be improved (because latencies are reduced with the ahead-scheduled transfers, which are basically a form of prefetching), but the energy consumption increases.[4] From a methodological point of view, Data Transfer and Storage Exploration [37] is one of several approaches that have been proposed to tackle this issue at different levels of abstraction, targeting either energy or performance improvements.

Most of these works studied, however, the problem of mapping static data types, which is very interesting because many structures in multimedia applications can be modeled as static structures and an appropriate mapping saves additional overheads at run-time. In contrast, this chapter proposes optimizations for heap data, allocated at run-time. Nevertheless, these approaches are compatible with this work and complement it in the overall endeavor of optimizing accesses to data allocated both at design and at run-time.

7.4.3 Dynamic Memory Management

As explained in Chap. 6, much research has also been performed [14, 27, 95, 112, 171] on the allocation techniques for dynamic data themselves, even trying to provide

---

[4] Energy consumption may increase in these cases because the cost of writing and reading input data from the scratchpad, and then writing and reading results before posting them to the DRAM, is added to the cost of accessing the data straight from the DRAM. As a result, independently of whether they are performed by the processor or the DMA, a net overhead of two writes and two reads to the scratchpad is created without any reutilization payback.

hardware support to accelerate the most common operations [7, 101]. However, these works do not consider the actual mapping of the pools into physical memory resources. Most of them focus instead on reducing the fragmentation of the heaps, while only the most recent explore options to increase performance or reduce energy consumption in embedded systems [14]. Actually, despite the fact that DDTs are becoming increasingly important as applications in embedded systems grow more complex and driven by external events [54], the heap has traditionally been left out of the assignment optimizations and mapped into the main DRAM.

Further evolution in the state of the art led to proposals such as [113], where the authors developed a specialized allocator for scratchpad memories. In that work, the internal data structures of the allocator are highly optimized using bitmaps to occupy as little memory as possible. That work is complementary to the proposal of this chapter because it provides mechanisms to manage a heap residing on a scratchpad memory, whereas here we propose an efficient placement of dynamic data into the system memories. In other words, the methodology proposed in this chapter studies the mapping of DDTs into heaps, and of heaps into the physical memory resources, while their work deals with the internal management of the heaps that have been assigned to a scratchpad memory.

A really interesting proposal to improve cache performance is presented in [94], where a compiler-based approach is used to build the "points-to" graph of the application DDTs and segregate every single instance of each DDT into a separate pool. However, this produces a worst-case assignment of pools to DDTs as the free space in a pool cannot be used to create instances of DDTs from other pools. Most importantly, that work was developed to improve the hit ratio of cache memories and hence it is not specifically suited for embedded systems with heterogeneous memory organizations.

Finally, a memory allocator for multi-threaded applications running on server-class multiprocessor systems is presented by [26]. Their allocator makes a careful separation of memory blocks into per-processor heaps and one global heap. However, their goal is not to take advantage of the memory organization, but to avoid the problems of false sharing (of cache lines) and incremented memory consumption under consumer-producer patterns in multiprocessor systems.

7.4.4 Placement

An early approximation to the problem of placement for dynamically allocated data is presented by [17], but the authors resort to a worst-case solution because they consider each allocation place in the source code as the declaration of a static variable and assign an upper bound on the amount of memory that can be used by each of them. Moreover, they consider each of these pseudo-static variables as independent entities, not taking into consideration the possibility of managing them in a single pool, and adding considerable complexity to their integer linear optimizer; in that sense, they lack a full approach to the concept of DM management. Nonetheless, that work constitutes one of the first approaches to the placement challenge. Further hints on how to map DDTs into a scratchpad memory are offered by [137]. In that work, the authors arrive to a satisfactory placement solution that reduces energy consumption and improves performance (albeit the latter only in multiprocessor systems) for a simple case; however, significant manual effort is still required from the designer.

In [105], we already presented a method that laid in the boundary between pure DM management, which deals with the problem of efficiently finding free blocks of memory for new objects, and the problem of placing dynamic objects into the right memory modules to reduce their access cost. There, we presented a method to build DMMs that can be configured to use a specific address range, but left open the way in which such address range is determined. For demonstration purposes, we manually found an object size that was receiving a big fraction of the application data accesses, created a pool specifically for it and mapped that pool into a separate scratchpad memory, obtaining important energy and time savings. The technique presented in this chapter is complementary because here we show how to find an appropriate placement of data into memory resources, while the actual mechanisms used to allocate blocks inside the pools are discussed in Chap. 6. Additionally, the methodology of this chapter classifies the allocations according not only to the size of the memory blocks, but also to the high-level data type of the objects.

More recently, efforts to create an algorithm to map dynamic, linked, data structures to a scratchpad memory have been presented by [57, 162], where the authors propose to place in the scratchpad some portions of the heap, called "*bins*". The bins are moved from the main memory to the scratchpad when the data they hold are known to be accessed at the next execution point of the application. For each DDT, a bin is created and only its first instances will reside in it (and so in the scratchpad), whereas the rest will be kept in a different pool permanently mapped into the main memory. That method requires a careful analysis of the application and instrumentation of the final code to execute the data movements properly. One drawback of this approach is that in order to offset the cost of the data movements, the application must spend enough time in the region of code that benefits from the new data distribution, reusing the data in the bins. Compared to this approach, our method avoids data migration between elements of the memory subsystem and considers the footprint of all the instances of the DDTs, not only the first ones.

The proposal of this chapter presents three important differences in relation with previous techniques. First, here we try to solve the problem of mapping all the DDTs, not only into one scratchpad memory, but into all the different elements of a heterogeneous memory subsystem. Second, we employ an exclusive memory organization model [83], which has some advantages under certain circumstances [158, 178]. In our case, this model avoids duplication of data across different levels of the memory subsystem; each level holds distinct data and no migrations are performed. Avoiding data movements reduces the energy and cycles overhead at the possible cost of using less efficiently the memory resources during specific phases of the application execution. However, due to our "liveness" analysis and the fact that no resources are wasted in duplicated data, we think that the methodology we have proposed in this chapter can overcome the possible inefficiencies. And third, this method can analyze the effect on the DDTs of the concurrent execution of all the application threads.

7.5 Conclusions

In this chapter, we have proposed a systematic approach to perform the placement of dynamic data types (DDTs) over the elements of a heterogeneous memory subsystem that tries to match the way in which each DDT is accessed with the different characteristics of each memory module. The grouping step reduces the memory footprint that would be required if each DDT were mapped over memory resources through its own independent memory pool, so that several DDTs can share the same memory resources without the need of data movements between elements in the memory hierarchy. The algorithms to manage the final pools are the ones presented in Chap. 6.

We have applied this approach to several application test cases. In the second case study we have evaluated the application of this methodology to the model of a wireless network routing application. In this case, we have seen that it is possible to obtain average improvements of 88.9% in energy consumption and 87.1% in the number of cycles of the memory subsystem when compared with systems using only LPDDR2 memories, and of 82% in energy consumption and 60.51% in cycles when compared with systems based on cache memories similar to that of an ARM Cortex-A15.

In an interesting opportunity for future improvements, the mapping process of this chapter could be executed at run-time, opening the possibility of adjusting the placement of any data object to cope with the changing conditions of the system, such as platform degradation along time, reduced availability of resources due to other concurrent applications or different models of the same platform with memory subsystems of different characteristics.

Finally, additional research could also be performed to optimize the cases where the instances of a given DDT have a very widely different number of accesses over the application execution period.

Appendix A
Description of Profiling Format

For reference purposes, Table A.1 introduces the different packets and packet fields that exist in the profiling format introduced in Chap. 3. The entries *AllocBegin* and *AllocEnd*, and *DeallocBegin* and *DeallocEnd*, are two profiling tokens that represent a single allocation or de-allocation event of the application. However, it is use full to split them because this allows to analyze the memory accesses that are performed during the (de)allocation process itself.

© Springer International Publishing Switzerland 2015
D. Atienza Alonso et al., *Dynamic Memory Management for Embedded Systems*,
DOI 10.1007/978-3-319-10572-7

Table A.1 Structure of the profiling packet for each specific event in the behavior of the application

Log packet	Field 1	Field 2	Field 3	Field 4	Field 5	Field 6	Field 7
Vector Construct	threadId	element Type	sequenceId	instanceId	element Size		
Vector Duplicate	threadId	element Type	sequenceId	instanceId	element Size	old SequenceId	old InstanceId
Vector Destruct	threadId	element Type	sequenceId	instanceId	element Size		
Vector Swap	threadId	element Type	sequenceId	instanceId	element Size	old SequenceId	old InstanceId
Vector Resize	threadId	element Type	sequenceId	instanceId	element Size		
Iterator Next	threadId	element Type	sequenceId	instanceId	element Size	address	
Iterator Previous	threadId	element Type	sequenceId	instanceId	element Size	address	
Iterator Add	threadId	element Type	sequenceId	instanceId	element Size	address	offset
Iterator Sub	threadId	element Type	sequenceId	instanceId	element Size	address	offset
Iterator Get	threadId	element Type	sequenceId	instanceId	element Size	address	
Vector Get	threadId	element Type	sequenceId	instanceId	element Size	index	
Vector Add	threadId	element Type	sequenceId	instanceId	element Size	index	
Vector Remove	threadId	element Type	sequenceId	instanceId	element Size	index	
Vector Clear	threadId	element Type	sequenceId	instanceId	element Size		
Vector End	threadId						
Var Read	threadId	VarId	address	size			
Var Write	threadId	VarId	address	size			

(continued)

Table A.1 (continued)

Log packet	Field 1	Field 2	Field 3	Field 4	Field 5	Field 6	Field 7
Alloc Begin	threadId	allocated Id	size				
Alloc End	threadId	allocated Id	size	address			
Dealloc Begin	threadId	allocated Id	address				
Dealloc End	threadId	allocated Id	address				
Scope Begin	threadId	scopeId					
Scope End	threadId	scopeId					
Thread Begin	threadId	oldThread Id					
Thread End	threadId						

References

1. M.J. Absar, F. Catthoor, Compiler-based approach for exploiting scratch-pad in presence of irregular array access, in DATE '05: Proceedings of the Conference on Design, Automation and Test in Europe (IEEE Computer Society, Washington, 2005), pp. 1162–1167
2. M.J. Absar, F. Poletti, P. Marchal, F. Catthoor, L. Benini Fast and power-efficient dynamic data-layout with DMA-capable memories, in *Proceedings of the International Workshop on Power-Aware Real-Time Computing (PACS)* (2004)
3. S. Agrawal, W. Thies, S. P. Amarasinghe, Optimizing stream programs using linear state space analysis, in CASES (2005), pp. 126–136
4. N. Ahmed, N. Mateev, K. Pingali, Tiling imperfectly-nested loop nests, in Proceedings of the 2000 ACM/IEEE Conference on Supercomputing (CDROM) (IEEE Computer Society, Washington, 2000).
5. A.V. Aho, R. Sethi, J.D. Ullman, *Compilers: Principles* (Techniques and Tools. Addison-Wesley Inc, Wokingham, 1986)
6. A. Alexandrescu, *Modern C++ Design: Generic Programming and Design Patterns Applied*, 1st edn. (Addison-Wesley Publishing Company Inc, Workingham, England, 2001)
7. I. Anagnostopoulos, S. Xydis, A. Bartzas, Z. Lu, D. Soudris, A. Jantsch, Custom microcoded dynamic memory management for distributed on-chip memory organizations. Embed. Syst. Lett. **3**(2), 66–69 (2011)
8. J.L. Antonakos, K.C Jr. Mansfield, Practical Data Structures using C/C++. (Prentice Hall, London, 1999).
9. ARM, Cortex-A15 Technical Reference Manual Rev. r2p0. ARM (2011).
10. D. Atienza, M. Leeman, J.M. Mendias, F. Catthoor, V. De Florio, G. Deconinck, Some experiences on dynamic memory management refinement at system-level for multimedia applications, in Proceedings of XVIII Conference on Design of Circuits and Integrated Systems, Ciudad Real, Spain (2003).
11. D. Atienza, S. Mamagkakis, F. Catthoor, J.M. Mendías, D. Soudris, Dynamic memory management design methodology for reduced memory footprint in multimedia and wireless network applications, in Proceedings of Design, Automation and Test in Europe Conference (DATE '04) (IEEE Press, Paris, France, 2004).
12. D. Atienza, S. Mamagkakis, F. Catthoor, J.M. Mendias, D. Soudris, Modular construction and power modelling of dynamic memory managers for embedded systems, in Proceedings of 14th Workshop on Power and Timing Modeling, Optimization and Simulation (PATMOS '04) Lecture Notes in Computer Science, vol. 3254 (Springer, Heidelberg, Santorini, Greece, 2004) pp. 510–520.

© Springer International Publishing Switzerland 2015
D. Atienza Alonso et al., *Dynamic Memory Management for Embedded Systems*,
DOI 10.1007/978-3-319-10572-7

13. D. Atienza, S. Mamagkakis, F. Catthoor, J.M. Mendias, D. Soudris. Reducing memory accesses with a system-level design methodology in customized dynamic memory management, in Proceedings of the 2nd Workshop on Embedded Systems for Real-Time Multimedia (ESTIMEDIA) (IEEE Computer Society, Stockholm, Sweden, 2004).

14. D. Atienza, S. Mamagkakis, F. Catthoor, J.M. Mendias, D. Soudris, Systematic dynamic memory management design methodology for reduced memory footprint. ACM Trans. Des. Autom. Embed. Syst. (TODAES) 11(2), 465–489 (2006)

15. D. Atienza, S. Mamagkakis, F. Poletti, J.M. Mendias, F. Catthoor, L. Benini, D. Soudris, Efficient system-level prototyping of power-aware dynamic memory managers for embedded systems. Integr. VLSI J. 39(2), 113–130 (2005)

16. D. Atienza et al., Optimization of dynamic data structures in multimedia embedded systems using evolutionary computation, in *Proceedings of 10th ACM International Workshop on Software & Compilers for Embedded Systems (SCOPES)* (Nice, France, 2007), pp. 31–40

17. O. Avissar, R. Barua, D. Stewart. Heterogeneous memory management for embedded systems, in Proceedings of the Fourth International Conference on Compilers, Architecture and Synthesis for Embedded Systems (CASES) (Atlanta, 2001).

18. C. Baloukas, J.L. Risco-Martin, D. Atienza, C. Poucet, L. Papadopoulos, S. Mamagkakis, D. Soudris, J. Ignacio Hidalgo, F. Catthoor, J. Lanchares. Optimization methodology of dynamic data structures based on genetic algorithms for multimedia embedded systems. J. Syst. Softw. 82(4), 590–602 (2009). (Special Issue: Selected papers from the 2008 IEEE Conference on Software Engineering Education and Training (CSEET08)) Engineering Education and Training (CSEET08).

19. R. Banakar, S. Steinke, B.-S. Lee, M. Balakrishnan, P. Marwedel, Scratchpad memory: A design alternative for cache on-chip memory in embedded systems, in Proceedings of the International Symposium on Hardware/Software Codesign (CODES) (ACM, New York, 2002), pp. 73–77.

20. F. Barat, M. Jayapala, T.V. Aa, R. Lauwereins, G. Deconinck, H. Corporaal, Low power coarse-grained reconfigurable instruction set processor, in Proceedings 13th International Workshop Field-Programmable Logic and Applications (FPL 2003) (2003).

21. A. Bartzas, S. Mamagkakis, G. Pouiklis, D. Atienza, F. Catthoor, D. Soudris, A. Thanailakis. Dynamic data type refinement methodology for systematic performance-energy design exploration of network applications. In DATE '06: Proceedings of the Conference on Design, Automation and test in Europe (European Design and Automation Association, Leuven, 2006), pp. 740–745 (3001).

22. A. Bartzas, M. Peon-Quiros, S. Mamagkakis, F. Catthoor, D. Soudris, J.M. Mendias, Enabling run-time memory data transfer optimizations at the system level with automated extraction of embedded software metadata information. In Proceedings of ASP-DAC (IEEE Computer Society Press, 2008), pp. 434–439.

23. N. Bellas, I.N. Hajj, C.D. Polychronopoulos, G. Stamoulis, Architectural and compiler techniques for energy reduction in high-performance microprocessors. IEEE Trans. Very Large Scale Integr. (VLSI). 8(3), 317–326 (2000).

24. L. Benini, A. Macii, E. Macii, M. Poncino, Increasing energy efficiency of embedded systems by application-specific memory hierarchy generation. IEEE Des. Test Comput 17(2), 74–85 (2000)

25. L. Benini, G. de Micheli, System-level power optimization: techniques and tools. ACM Trans. Des. Autom. Embed. Syst. 5(2), 115–192 (2000)

26. E.D. Berger, K.S. McKinley, R.D. Blumofe, P.R. Wilson, Hoard: A scalable memory allocator for multithreaded applications. SIGPLAN Not. 35(11), 117–128 (2000)

27. E.D. Berger, B.G. Zorn, K.S. McKinley, Composing high-performance memory allocators. In Proceedings ACM SIGPLAN Conference on Programming Language Design and Implementation (PLDI) (Snowbird, Utah, 2001), pp. 114–124.

28. S.M. Blackburn, K.S. McKinley, Ulterior reference counting: Fast garbage collection without a long wait. In Proceedings of SIGPLAN 2003 Conference on Object-Oriented Programming, Systems, Languages and Applications, OOPSLA'03, Anaheim, California, USA, October, 2003 (ACM SIGPLAN Notices. ACM Press, 2003).

29. Boost.org. Boost c++ libraries (2009), http://www.boost.org/
30. D.P. Bovet, M. Cesati. Understanding the Linux Kernel, From I/O Ports to Process Management (O'Reilly and Associates, Sebastopol, 2001) (1001 Morris Street).
31. G. Brassard, T. Bratley, Fundamentals of Algorithmics, 1st edn. (Prentice Hall, Englewood, 1996), pp. 227–230 (Spanish).
32. D. Bulka, D. Mayhew, *Efficient C++* (Addison-Wesley, Wokingham, 2001)
33. C++ Standardisation Comittee. Programming languages - C++ - ISO/IEC 14882. Technical report, American National Standards Institutes, 11 West 42nd Street, New York, New York 10036, USA, (1998).
34. F. Catthoor, E. Brockmeyer, *Unified Meta-Flow Summary for Low-Power Data-Dominated Applications, Chapter in Unified Low-Power Design Flow for Data-Dominated Multi-Media and Telecom Applications* (Kluwer Academic Publishers, Boston, USA, 2000)
35. F. Catthoor, K. Danckaert, C. Kulkarni, E. Brockmeyer, P.G. Kjeldsberg, T. Van Achteren, T. Omnes, *Data Access and Storage Management for Embedded Programmable Processors* (Kluwer Academic Publishers, Boston, 2002)
36. F. Catthoor, P. Raghavan, A. Lambrechts, M. Jayapala, A. Kritikakou, J. Absar, *Ultra-Low Energy Domain-Specific Instruction-Set Processors* (Springer, Germany, 2010)
37. F. Catthoor, S. Wuytack, E. De Greef, F. Balasa, L. Nachtergaele, A. Vandecappelle, Custom Memory Management Methodology. Exploration of Memory Organisation for Embedded Multimedia System Design (Kluwer Academic Publishers, Boston, 1998).
38. J. Morris Chang, C.-T. Dan Lo, W. Srisa-an, Omx: object management extension. In The Second International Workshop on Compiler and Architecture Support for Embedded Systems (CASES) (San Diego, 1999).
39. C.C. Shan, Shift to control. In 2004 Scheme Workshop (2004).
40. J. Cockx, K. Denolf, B. Vanhoof, R. Stahl, SPRINT: a tool to generate concurrent transaction-level models from sequential code. EURASIP J. Appl. Sig. Process. **2007**(1), 213–213 (2007)
41. Carlos A. Coello, David A. Van Veldhuizen, Gary B. Lamont, *Evolutionary Algorithms for Solving Multi-Objective Problems* (Kluwer Academic Publishers, New York, 2002)
42. Thomas H. Cormen, Charles E. Leiserson, Ronald L. Rivest, Clifford Stein, *Introduction to Algorithms*, 3rd edn. (MIT Press, USA, 2009)
43. Microsoft Corporation: Heap:pleasures and pains (for Windows NT Technologies). http://msdn.microsoft.com/library/default.asp?url=/library/en-us/dngenlib/html/heap3.asp
44. Microsoft Corporation: Heaps in Windows CE. http://msdn.microsoft.com/library/default.asp?url=/library/en-us/wcecoreos5/html/wce50conheaps.asp
45. J. Cosmas, I. Taki, D. Green, O. Zalesny, L. Van Gool, M. Pollefeys, R. Degeest, M. Waelkens, K. Hraby, M. Kampel, R. Sablatnig. 3D Murale (2002), http://www.brunel.ac.uk/project/murale/home.html
46. D. Coutts, R. Leshchinskiy, D. Stewart, Stream fusion: From lists to streams to nothing at all. In Proceedings of the ACM SIGPLAN International Conference on Functional Programming, ICFP (2007).
47. C. Poucet, D. Atienza, F. Catthoor, Template-based semi-automatic profiling of multimedia applications. In Proceedings of the International Conference on Multimedia and Expo (ICME) (IEEE Computer, Signal Processing, System and Communications Society, Toronto, Canada, 2006). pp. 1061–1064.
48. O. Danvy, A. Filinski, Abstracting control. In Proceedings of the 1990 ACM Conference on LISP and Functional Programming, Nice (ACM, New York, 1990), pp. 151–160.
49. O. Danvy, A. Filinski, Representing control: a study of the CPS transformation. Math. Struct. Comput. Sci. **2**(4), 361–391 (1992)
50. M. Dasygenis, E. Brockmeyer, B. Durinck, F. Catthoor, D. Soudris, A. Thanailakis, A combined DMA and application-specific prefetching approach for tackling the memory latency bottleneck. IEEE Trans. Very Large Scale Integr. Syst. (TVLSI) 14(3), 279–291 (2006).
51. P. Cheng, D.F. Bacon, V.T. Rajan, A real-time garbage collector with low overhead and consistent utilization. In Proceedings of SIGPLAN 2003 Symposium on Principles of Programming Languages, POPL'03, New Orleans, Lousiana, USA (ACM SIGPLAN Notices, ACM Press, 2003).

52. E.G. Daylight, D. Atienza, A. Vandecappelle, F. Catthoor, J.M. Mendias, Memory-access-aware data structure transformations for embedded software with dynamic data accesses. IEEE Trans. VLSI Syst. **12**, 269–280 (2004)

53. E. De Greef, Storage size reduction for multimedia applications. Ph.D. thesis, Fakulteit Toegepaste Wetenschappen, Katholieke Universiteit Leuven, 1998.

54. H. De Man, Connecting E-Dreams to Deep-Submicron Realities (Springer, Berlin, 2004) in *Proceedings of Workshop on Power and Timing Modeling, Optimization and Simulation PATMOS '04* (Santorini, Greece, 2004).

55. G. de Micheli, L. Benin (eds.), *Networks on Chips* (Technology and Tools (Morgan Kaufmann Publishers, San Francisco, 2006)

56. S. Debray, W. Evans, R. Muth, B. De Sutter, Compiler techniques for code compaction. ACM Trans. Prog. Lang. Syst. **22**(2), 378–415 (2000)

57. A. Dominguez, S. Udayakumaran, R. Barua, Heap data allocation to scratch-pad memory in embedded systems. J. Embed. Comput. **1**(4), 521–540 (2005)

58. L. Eeckhout, A. Georges, K. De Bosschere, How java programs interact with virtual machines at the microarchitectural level. In Proceedings of the ACM Conference on Object-Oriented Programming, Systems, Languages, and Applications (OOPSLA) (ACM Press New York, Anaheim, 2003).

59. I. Sodagar et al., Scalable wavelet coding for synthetic and natural hybrid images. IEEE Trans. Circ. Syst. Video Technologhy **9**(2), 353–369 (1999)

60. Eyetronics 3d scanning solutions. http://www.eyetronics.com

61. E. Fredkin, Trie memory. Commun. ACM **3**(9), 490–499 (1960)

62. i.mx287 multimedia applications processors (2011), http://www.freescale.com/imx287

63. B.S. Furber, *ARM System-on-Chip Architecture* (Addison-Wesley Longman Publishing Co., Inc, Boston, 2000)

64. E. Gamma, R. Helm, R. Johnson, J. Vlissides, *Design Patterns: Elements of Reusable Object-Oriented Software* (Addison-Wesley Longman Publishing Co., Inc, Boston, 1995)

65. M. Gasbichler, M. Sperber, Final Shift for Call/cc: Direct Implementation of Shift and Reset (2002).

66. D. Gay, A. Aiken, Memory management with explicit regions. In Proceedings of ACM SIGPLAN 2001 Conference on Programming Language Design and Implementation (PLDI '01) (Snowbird, Utah, 2001), pp. 70–80.

67. B. Geelen, E. Brockmeyer, B. Durinck, G. Lafruit, R. Lauwereins, Alleviating memory bottlenecks by software-controlled data transfers in a data-parallel wavelet transform on a multicore DSP. in Proceedings of the IEEE BENELUX/DSP Valley Signal Processing Symposium (SPS-DARTS) (2005), pp. 143–146.

68. S.V. Gheorghita, M. Palkovic, J. Hamers, A. Vandecappelle, S. Mamagkakis, T. Basten, L. Eeckhout, H. Corporaal, F. Catthoor, F. Vandeputte, K. De Bosschere, System-scenario-based design of dynamic embedded systems. ACM Trans. Des. Automa. Electron. Syst. (TODAES) **14**(1), 1–45 (2009)

69. A. Gill, J. Launchbury, S.L. Peyton Jones, A short cut to deforestation. in Conference on Functional Programming Languages and Computer Architecture (1993), pp. 223–232.

70. R. González-Alberquilla, F. Castro, L. Piñuel, F. Tirado, Stack filter: Reducing L1 data cache power consumption. J. Syst. Architect. **56**, 685–695 (2010)

71. Handheld quake. http://handheldquake.sourceforge.net/

72. C.J. Harris, M. Stephens, A combined corner and edge detector. in 4th Alvey Vision Conference (Manchester, 1988), pp. 147–151.

73. T. Henderson, D. Kotz, I. Abyzov, The changing usage of a mature campus-wide wireless network. in Proceedings of the International Conference on Mobile Computing and Networking (MobiCom) (ACM, New York, 2004), pp. 187–201.

74. J.L. Hennessy, D.A. Patterson, *Computer architecture: a quantitative approach*, 5th edn. (Morgan Kaufmann, San Francisco, CA, USA, 2011)

75. J. Holland, *Adaptation in Natural and Artificial Systems* (University of Mitchigan Press, Ann Arbor, 1975)

76. CACTI 5.3 (2008), http://quid.hpl.hp.com:9081/cacti/
77. CACTI 6.0 (2011), http://www.hpl.hp.com/research/cacti/
78. Sun Microsystems Inc., The collections framework (2005), http://java.sun.com/docs/books/tutorial/collections/
79. F. Ingelrest, G. Barrenetxea, G. Schaefer, M. Vetterli, O. Couach, M. Parlange, SensorScope: application-specific sensor network for environmental monitoring. ACM TOSN **6**(2), 1–32 (2010)
80. K. Ishibashi, K. Osada (eds.), *Low Power and Reliable SRAM Memory Cell and Array Design* (Springer, Germany, 2011)
81. JEDEC: Low Power Double Data Rate 2 (LPDDR2)–JESD209-2E. JEDEC Solid State Technology Association (2011).
82. M.N. Josuttis, *The C++ Standard Library* (Addison Wesley, Harlow, 1999)
83. N. Jouppi, S. Wilton, Tradeoffs in two-level on-chip caching. in Proceedings of the IEEE ISCA (1994), pp. 34–45.
84. N. Jouppi, Western research laboratory, cacti (2002), http://research.compaq.com/wrl/people/jouppi/CACTI.html
85. M. Kandemir, J. Ramanujam, A. Choudhary, Improving cache locality by a combination of loop and data transformations. IEEE Trans. Comput. **48**(2), 159–167 (1999)
86. M. Kandemir, J. Ramanujam, M.J. Irwin, N. Vijaykrishnan, I. Kadayif, A. Parikh, Dynamic management of scratch-pad memory space, in Proceedings of Design Automation Conference (DAC) (Las Vegas, USA, 2001), pp. 690–695.
87. M. Kandemir, I. Kadayif, A. Choudhary, J. Ramanujam, I. Kolcu. Compiler-directed scratch-pad memory optimization for embedded multiprocessors. IEEE Trans. Very Large Scale Integr. Syst. (TVLSI) 12, 281–287 (2004).
88. L. Kharevych, R. Khan, *3D Physics Engine for Elastic and Deformable Bodies* (University of Maryland, College Park , 2002)
89. O. Kiselyov, General ways to traverse collections (2000).
90. D. Kotz, K. Essien, Analysis of a campus-wide wireless network. Wireless Netw. **11**(1–2), 115–133 (2005)
91. J. Kozubik, Freebsd and solid state devices (2001), http://www.freebsd.org/doc/en_US.ISO8859-1/articles/solid-state/
92. Lawrence Berkeley National Laboratory, The internet traffic archive (2000), http://ita.ee.lbl.gov/
93. W. Landi, B.G. Ryder, S. Zhang, Interprocedural modification side effect analysis with pointer aliasing, in pldi93 (Albuquerque, 1993), pp. 56–67.
94. C. Lattner, V. Adve, Automatic pool allocation: Improving performance by controlling data structure layout in the heap, in Proceedings of PLDI (ACM, 2005), pp. 129–142.
95. D. Lea, A memory allocator (1996), http://g.oswego.edu/dl/html/malloc.html
96. D. Lea, The lea 2.7.2 dynamic memory allocator (2002), http://gee.cs.oswego.edu/dl/
97. M. Leeman, D. Atienza, F. Catthoor, G. Deconinck, J.M. Mendias, V. De Florio, R. Lauwereins, Intermediate variable elimination in a global context for a 3D multimedia application, in Proceedings of International Conference on Multimedia and Expo (ICME) (IEEE Computer Society, Signal Processing Society, System, Baltimore, 2003)
98. M. Leeman, D. Atienza, F. Catthoor, G. Deconinck, J.M. Mendias, V. De Florio, R. Lauwereins, Power estimation approach of dynamic data storage on a hardware software boundary level, in Integrated Circuits and System Design-Power And Timing Modeling, Optimization and Simulation, 13th International, Workshop, PATMOS 2003, eds. by J.J. Chico, E. Macii. Lecture Notes in Computer Science, vol. 2799 (Springer, Heidelberg, Turin, Italy, 2003), pp. 289–298.
99. M. Leeman, D. Atienza, G. Deconinck, V. Florio, J.M. Mendías, C. Ykman-Couvreur, F. Catthoor, R. Lauwereins, Methodology for refinement and optimisation of dynamic memory management for embedded systems in multimedia applications. J. VLSI Sig. Process. Syst. **40**(3), 383–396 (2005)

100. M. Leeman, Interactive Strategies and Analysis Method for Dynamic Data Type Transformation and Refinement in Multimedia Applications, Ph.D. thesis, Katholieke Universiteit Leuven, 2003.

101. W. Li, S. Mohanty, K. Kavi, A page-based hybrid (software-hardware) dynamic memory allocator. IEEE. Comput. Archit. Lett. **5**(2), 13–13 (2006)

102. A.W. Lim, S.-W. Liao, M.S. Lam, Blocking and array contraction across arbitrarily nested loops using affine partitioning, in Proceedings of the ACM SIGPLAN Symposium on Principles and Practices of Parallel Programming (PPoPP) (ACM, New York, NY, USA, 2001), pp. 103–112.

103. M. Loghi, F. Angiolini, D. Bertozzi, L. Benini, R. Zafalon, Analyzing on-chip communication in a mpsoc environment, *Proceedings of the Conference on Design, Automation and Test in Europe (IEEE Computer Society* (France, Paris, 2004), p. 20752

104. D. Luebke, M. Reddy, J. Cohen, A. Varshney, B. Watson, R. Huebner, *Level of Detail for 3D Graphics* (Morgan-Kaufmann Publishers, San Francisco, 2002)

105. S. Mamagkakis, D. Atienza, C. Poucet, F. Catthoor, D. Soudris, Energy-efficient dynamic memory allocators at the middleware level of embedded systems, in Proceedings of EMSOFT (ACM, 2006), pp. 215–222.

106. S. Mamagkakis, D. Atienza, C. Poucet, F. Catthoor, D. Soudris, J.M. Mendias, Custom design of multi-level dynamic memory management subsystem for embedded systems, in Proceedings of the IEEE Workshop on Signal Processing Systems (SIPS) (IEEE Press, Austin, Texas, USA, 2004), pp. 170–175.

107. S. Mamagkakis, C. Baloukas, D. Atienza, F. Catthoor, D. Soudris, A. Thanailakis, Reducing memory fragmentation in network applications with dynamic memory allocators optimized for performance. Comput. Commun. **29**(13–14), 2612–2620 (2006)

108. S. Mamagkakis, A. Bartzas, G. Pouiklis, D. Atienza, F. Catthoor, D. Soudris, A. Thanailakis, Systematic methodology for exploration of performance–energy trade-offs in network applications using dynamic data type refinement. J. Syst. Architect. **53**(7), 417–436 (2007)

109. P. Marchal, F. Catthoor, D. Bruni, L. Benini, J.I. Gómez, L. Piñuel, Integrated task scheduling and data assignment for SDRAMs in dynamic applications. IEEE Des. Test Comput. **21**(5), 378–387 (2004)

110. P. Marchal, J.I. Gómez, F. Catthoor, Optimizing the memory bandwidth with loop fusion. in CODES+ISSS '04: Proceedings of the 2nd IEEE/ACM/IFIP International Conference on Hardware/Software Codesign and System Synthesis (ACM, New York, NY, USA, 2004), pp. 188–193.

111. P. Marchal, C. Wong, A. Prayati, N. Cossement, F. Catthoor, R. Lauwereins, D. Verkest, H. De Man, Impact of task level concurrency transformations on the mpeg4 im1 player for weakly parallel processor platform, in Proceedings of Parallel Architectures and Compilation Techniques (PACT) (Philadelphia PA, USA, October 2003).

112. B.H. Margolin, R.P. Parmelee, M. Schatzoff, Analysis of free-storage algorithms. IBM Syst. J. **10**(4), 283–304 (1971)

113. R. McIlroy, P. Dickman, J. Sventek, Efficient dynamic heap allocation of scratch-pad memory, in Proceedings of ISMM (ACM, 2008), pp. 31–40.

114. E. Meijer, M. Fokkinga, R. Paterson, in Proceedings 5th ACM Conf. on Functional Programming Languages and Computer Architecture, FPCA'91, ed by J. Hughes. Functional programming with bananas, lenses, envelopes and barbed wire, vol. 523 (Springer-Verlag, Berlin, Cambridge, MA, USA, 1991) 26–30 Aug 1991, pp. 124–144.

115. G. Memik, B. Mangione-Smith, W. Hu, Netbench: a benchmarking suite for network processors. Technical report, CARES Technical Report 2001-2-01 (2001).

116. G. Memik, W.H. Mangione-Smith, W. Hu, Netbench: a benchmarking suite for network processors, in ICCAD '01: Proceedings of the 2001 IEEE/ACM International Conference on Computer-Aided Design (IEEE Press, Piscataway, NJ, USA, 2001), pp. 39–42.

117. T.H. Meng, B. Gordon, E. Tsern, A. Hung, Portable video-on-demand in wireless communication. Proc. IEEE Spec. Issue Low Power Electron. **83**(4), 659–680 (1995).

118. Z. Michalewicz, *Genetic Algorithms + Data Structures = Evolution Programs* (Springer-Verlag, Berlin, 1996)
119. MICRON, Mobile LPSDR SDRAM–MT48H32M32LF/LG Rev. D 1/11 EN (Micron Technology Inc, 2010).
120. MICRON, Mobile LPDDR2 SDRAM–MT42L64M32D1 Rev. N 3/12 EN (Micron Technology Inc, 2012).
121. J. Middleton, Voice over ip: setting phone service free. IEEE Spectr. **16**, 100–110 (2010).
122. Moby games, a game documentation and review project, http://www.mobygames.com/
123. T. Möller, E. Haines, *Real-Time Rendering* (A. K. Peters Ltd, Natick, 1999)
124. ISO/IEC JTC1/SC29/WG11 MPEG-4 standard features overview, http://www.chiariglione.org/mpeg/standards/mpeg-4/mpeg-4.htm
125. Microsoft MSDN, Low fragmentation heap in XP technologies, http://msdn.microsoft.com/library/default.asp?url=/library/en-us/memory/base/low_fragmentation_heap.asp
126. S. Muchnick, *Advanced Compiler Design and Implementation* (Morgan Kaufmann Publisher, San Francisco, 1997) (April 2000)
127. N. Murphy, Safe memory usage with dynamic memory allocation. Embed. Syst. 49–57 (2000).
128. On-Line Application Research (OAR). Rtems, open-source real-time operating system for multiprocessor systems (2002), http://www.rtems.org
129. University of Glasgow. Glasgow haskell compiler (1992), http://www.haskell.org/ghc/
130. OMG: Unified modeling language (uml) (2014).
131. A. Osyczka, in Multicriteria Optimization for Engineering Design, ed by J.S. Gero. Design Optimization (Academic Press, New Jersey, 1985) pp. 193–227.
132. P.R. Panda, F. Catthoor, N.D. Dutt, K. Danckaert, E. Brockmeyer, C. Kulkarni, Data and memory optimizations for embedded systems. ACM Trans. Design Autom. Embedded Syst. **6**(2), 142–206 (2001)
133. P.R. Panda, N.D. Dutt, A. Nicolau, On-chip vs. off-chip memory: The data partitioning problem in embedded processor-based systems. ACM Trans. Design Autom. Electron. Syst. **5**(3), 682–704 (2000)
134. D.A. Patterson, J.L. Hennessy, *in Computer Architecture: A Quantitative Approach* , 1st edn. (Morgan Kaufmann, San Mateo, 1990)
135. M. Peon-Quiros, A. Bartzas, S. Mamagkakis, F. Catthoor, J.M. Mendias, D. Soudris, *Direct Memory Access Optimization in Wireless Terminals for Reduced Memory Latency and Energy Consumption* (Springer, in *Proceedings of Workshop on Power and Timing Modeling, Optimization and Simulation—PATMOS, 2007)*
136. N.P. Ngoc, W. van Raemdonck, G. Lafruit, G. Deconinck, R. Lauwereins, Qos framework for interactive 3d applications, in Proceedings of 10th International Conference in Central Europe on Computer Graphics, Visualization and Computer Vision (Plzen, Czechoslovakia, 2002), pp. 317–325.
137. F. Poletti, P. Marchal, D. Atienza, L. Benini, F. Catthoor, J.M. Mendias, An integrated hardware/software approach for run-time scratch-management, in Accepted for Proceedings of 41st Design Automation Conference (San Diego, USA, 2004).
138. M. Pollefeys, R. Koch, M. Vergauwen, L. Van Gool, Metric 3D surface econstruction from uncalibrated image sequences, in Lecture Notes in Computer Science. Proceedings SMILE Workshop (post-ECCV'98) vol. 1506, (Springer, 1998), pp. 139–153.
139. C. Poucet, S. Mamagkakis, D. Atienza, and F. Catthoor. Systematic intermediate sequence removal for reduced memory accesses. In Proceedings of the 10th International Workshop on Software and Compilers for Embedded Systems (SCOPES 2007), Nice, France, 2007.
140. F. Quillere, S. Rajopadhye, Optimizing memory usage in the polyhedral model. ACM Trans. Program. Lang. Syst. **22**(5), 773–815 (2000)
141. Samsung. Exynos 4 quad, 32nm hkmg process (2014), http://www.samsung.com/global/business/semiconductor/minisite/Exynos/w/solution.html#?v=quad
142. A. Santos, Compilation by Transformation in Non-Strict Functional Languages. Ph.D. thesis (1995).

143. J.D. Schaffer, Multiple objective optimization with vector evaluated genetic algorithms, in Proceedings of the 1st International Conference on Genetic Algorithms (Lawrence Erlbaum Associates Inc., Mahwah, 1985), pp. 93–100.

144. N. Schemenauer, T. Peters, M. Lie Hetland, Pep 255: Simple generators, Python Enhancement Proposals (2006).

145. SGI. Standard template library, 2006. http://www.sgi.com/tech/stl/

146. M. Shalan, V.J. Mooney II, A dynamic memory management unit for embedded real-time system-on-a-chip, in Proceedings of the third International Workshop on Compiler and Architecture Support for Embedded Systems (CASES) (San Jose, California, USA, 2000).

147. M. Shreedhar, G. Varghese, *Efficient fair queuing using deficit round robin* (In In Proc. of SIGCOMM, Cambridge, MA, 1995)

148. M. Shreedhar, G. Varghese, Efficient fair queueing using deficit round-robin. IEEE/ACM Trans. Networking **4**(3), 375–385 (1996)

149. A. Smailagic, D.P. Siewiorec, D. Anderson, C. Kasaback, T. Martin, J. Stivoric. Benchmarking an interdisciplinary concurrent design methodology for electronic/mechanical systems. In: Proceedings of the 32nd ACM/IEEE conference on Design Automation Conference (DAC) (ACM Press, New York, NY, 1995), pp. 514–519.

150. Y. Smaragdakis, D. Batory, Mixin-based programming in C++. Lect. Notes Comput. Sci. **2177**, 163 (2001)

151. M. Soto, A. Rossi, M. Sevaux, A mathematical model and a metaheuristic approach for a memory allocation problem. J. Heuristics **18**(1), 149–167 (2012)

152. Sourceforge. Simblob–the 3d environment builder framework. http://sourceforge.net/projects/simblob

153. Sourceforge. Vdrift racing simulator. http://sourceforge.net/projects/vdrift

154. W. Srisa-an, C.-T. Dan, Lo, J.M. Chang. Active memory processor: A hardware garbage collector for real-time java embedded devices. IEEE Trans. Mobile Comput. **2**(2), 89–101 (2003)

155. S. Steinke, N. Grunwald, L. Wehmeyer, R. Banakar, M. Balakrishnan, P. Marwedel. Reducing energy consumption by dynamic copying of instructions onto onchip memory. In: Proceedings IEEE/ACM International Symposium on System Synthesis (ISSS) (IEEE Press, 2002).

156. S. Steinke, L. Wehmeyer, B. Lee, P. Marwedel, Assigning program and data objects to scratch-pad for energy reduction, in: Proceedings of Design, Automation and Test in Europe (DATE) (2002), p. 409.

157. A.A. Stepanov M. Lee, The Standard Template Library. Technical Report X3J16/94-0095, WG21/N0482 (1994).

158. S. Subha, *An exclusive cache model, in IEEE ITNG* (NV, Las Vegas, 2009), pp. 1715–1716

159. A.S. Tanenbaum, D.J. Wetherall, *Computer Networks*, 5th edn. (Prentice Hall, Englewood Cliffs, 2010)

160. Target jr (2002), http://www.robots.ox.ac.uk/tgtjr/

161. Ti's omap platform (2004), http://focus.ti.com/omap/docs/

162. S. Udayakumaran, A. Dominguez, R. Barua, Dynamic allocation for scratch-pad memory using compile-time decisions. ACM TECS **5**(2), 472–511 (2006)

163. F. Vahid, Procedure exlining: a new system-level specification transformation, in EURO-DAC '95/EURO-VHDL '95: Proceedings of the Conference on European design automation (IEEE Computer Society Press, Los Alamitos, 1995), pp. 508–513.

164. D. van Arkel, J. van Groningen, and S. Smetsers. Fusion in practice, in In Implementation of Functional Languages, (LNCS) (2002), pp. 51–67.

165. P. Vanbroekhoven, H. Corporaal, G. Janssens, F. Catthoor, M. Bruynooghe, Advanced copy propagation for arrays, in Proceedings of Languages, Compilers and Tools for Embedded Systems (San Diego CA, USA, 2003).

166. D. Vandevoorde, N.M. Josuttis, *C++ Templates The Complete Guide* (Addison Wesley, London, UK, 2003)

167. M. Verma, S. Steinke, P. Marwedel, Data partitioning for maximal scratchpad usage, in Proceedings of the Asia and South Pacific Design Automation Conference (ASP-DAC) (2003), pp. 77–83.

168. M. Verma, L. Wehmeyer, P. Marwedel, Cache-aware scratchpad allocation algorithm, in Proceedings of Design, Automation and Test in Europe (DATE) (2004).

169. K.-P. Vo, Vmalloc: s general and efficient memory allocator. Softw. Pract Experience **26**, 1–18 (1996)

170. P. Wadler, Deforestation: Transforming programs to eliminate trees, in ESOP '88. European Symposium on Programming, Nancy, France. Lecture Notes in Computer Science, vol. 300 (Springer-Verlag, Berlin, 1988), pp. 344–358.

171. P.R. Wilson, M.S. Johnstone, M. Neely, D. Bowles, *Dynamic storage allocation: a survey and critical review, in International Workshop on Memory Management* (Kincross, Scotland, UK, Springer Verlang LNCS, 1995), pp. 1–116

172. M. Woo, J. Neider et al., *OpenGL Programming Guide*, 2nd edn. (Silicon Graphics Inc, USA, 1997)

173. Derick Wood, *Data structures, Algorithms, and Performance* (Addison-Wesley Longman Publishing Co., USA, 1993)

174. S. Wuytack, F. Catthoor, H. De Man, Transforming set data types to power optimal data structures. IEEE Trans. Comput. Aided Des **15**(6), 619–629 (1996)

175. S. Wuytack, F. Catthoor, F. Franssen, L. Nachtergaele, H. De Man, Global communication and memory optimizing transformations for low power systems, in IEEE International Workshop on Low Power Design (Napa, 1994), pp. 203–208.

176. S. Wuytack, J.-P. Diguet, F. Catthoor, H. De Man, Formalized methodology for data reuse exploration for low-power hierarchical memory mappings. IEEE Trans. Very Large Scale Integr. Syst. (TVLSI) 6(4), 529–537 (1998).

177. S. Wuytack, J.P. Diguet, F. Catthoor, H. De Man, Formalized methodology for data reuse exploration for low-power hierarchical memory mappings. IEEE Trans. VLSI Syst. **6**(4), 529–537 (1998)

178. Y. Zheng, B.T. Davis, M. Jordan, Performance evaluation of exclusive cache hierarchies, in Proceedings of IEEE ISPASS (Washington, DC, USA, 2004), pp. 89–96.